ADVANCES IN CELL AGING AND GERONTOLOGY

VOLUME 9

Stem Cells:
A Cellular Fountain
of Youth

ADVANCES IN CELL AGING AND GERONTOLOGY

VOLUME 9

Stem Cells:
A Cellular Fountain
of Youth

Edited by

MARK P. MATTSON

Laboratory of Neurosciences
National Institute on Aging
Baltimore, MD 21224
USA

and

GARY VAN ZANT

Department of Hematology/Oncology
University of Kentucky Medical Center
Lexington, KY 40536-0093
USA

2002

ELSEVIER

AMSTERDAM – BOSTON – LONDON – NEW YORK – OXFORD – PARIS
SAN DIEGO – SAN FRANCISCO – SINGAPORE – SYDNEY – TOKYO

ELSEVIER SCIENCE B.V.
Sara Burgerhartstraat 25
P.O. Box 211, 1000 AE Amsterdam, The Netherlands

First edition 2002

British Library Cataloguing in Publication Data

```
Stem cells : a cellular fountain of youth?. - (Advances in
   cell aging and gerontology ; 9)
   1.Cells - Aging 2.Gerontology 3.Stem cells
   I.Mattson, Mark Paul II.Van Zant, Gary
   612.6'7

   ISBN 0444507310
```

Library of Congress Cataloging in Publication Data
A catalog record from the Library of Congress has been applied for.

ISBN: 0-444-50731-0
ISSN: 1566-3124   (Series)

♾ The paper used in this publication meets the requirements of ANSI/NISO Z39.48-1992 (Permanence of Paper).

Printed and bound by CPI Group (UK) Ltd, Croydon, CR0 4YY

Transferred to digital print 2012

TABLE OF CONTENTS

PREFACE

Stem cells can be considered the antithesis of aging and, indeed, stem cells are defined by their immortal self-renewing properties. The fertilized egg is the ultimate stem cell, being the predecessor of the hundreds of billions of cells that comprise the body. During very early embryonic development, individual cells in the blastocyst (embryonic stem cells) retain their ability to generate the entire body. At later times in development, multi-potent stem cells become more restricted and eventually each tissue harbors populations of progenitor cells that have the capability of forming mainly the cell types that comprise that specific tissue. For example, in the nervous system, neural progenitor cells give rise to glia and neurons, whereas in the bone marrow, stem cells give rise to the many different types of blood cells including lymphocytes and macrophages. The present volume of *Advances in Cell Aging and Gerontology* covers the cellular and molecular mechanisms regulating stem cell fate in relation to aging and age-related disease.

The book begins with a chapter by Thomas Smithgall, which describes the fundamental properties of embryonic stem cells with a focus on mechanisms regulating their self-renewal and differentiation capabilities. It is human embryonic stem cells that are the subject of current controversy from the standpoint of the ethics of using these cells to generate new organs and to treat various diseases. It is already becoming clear that embryonic stem cells have considerable potential for replacing dysfunctional or dead cells in an array of age-related diseases including diabetes, cardiovascular disease and neurodegenerative disorders.

The next three chapters focus on bone marrow stem cells. Gary Van Zant and colleagues discuss intriguing findings suggesting that there is genetic control on the self-renewing capability of bone marrow, and that there is a relationship between lifespan of the species and the functional capabilities of hematopoietic stem cells. Amelia Globerson discusses the role of hematopoietic stem cells in the aging thymus. It has long been known that the size of the thymus progressively decreases with aging, and an understanding of the stem cell populations that have the potential to re-populate the thymus is of considerable interest in understanding immunological decline during aging. Karen Chandross and Eva Mezey discuss the exciting findings concerning the possibility that progenitor cells in one tissue in the adult are capable of forming cells of other tissue types when placed in the proper environment. They focus on the plasticity of adult bone marrow stem cells and the signaling mechanisms that regulate their differentiation into either blood cells or cells of other tissue such as the brain.

Two chapters discuss aging and neural stem cells. It is now clear that the adult brain contains populations of stem cells capable of dividing and differentiating into neurons, astrocytes and oligodendrocytes. An increasing knowledge of the molecular regulation of these neural stem cells is leading to an understanding of their role in adult brain plasticity, for example, in learning and memory. In addition, pre-clinical studies and in some cases clinical studies are in progress to establish whether transplantation of embryonic or neural stem cells into patients with disorders such as Parkinson's disease and stroke can be effective in restoring brain function. Jingli Cai and Mahendra Rao discuss molecular and cellular mechanisms regulating neural stem cell fate, and my colleagues and I discuss neural stem cells in the context of neurodegenerative disorders such as Alzheimer's disease, Parkinson's disease and Huntington's disease.

Ken Boheler and Anna Wobus present a chapter describing exciting findings on the role of stem cells in myocardial development and aging. There have rapid advances in understanding of the regulation of heart development and in developing stem cell-based therapies for repairing damaged heart tissue and promoting re-growth of blood vessels. Adult skeletal muscle contains populations of stem cells that likely play a major role in regeneration of damaged muscle. Patrick Seale and Michael Rudnicki discuss basic aspects of the cell and molecular biology of skeletal muscle regeneration, and exciting findings suggesting that myogenic stem cells also have the potential to differentiate into other cell types and can even reconstitute the hematopoietic system when administered intravenously. These findings further emphasize the plasticity of stem cells in the adult. In the final chapter, Robert Jilka and Michael Parfitt provide an overview of the role of stem cells in bone formation and remodeling in the adult. A decline in bone strength is a major problem among the elderly and an understanding of how stem cells contribute to bone development may provide new avenues for the prevention and treatment of bone disorders of aging.

Collectively the chapters in this book provide the reader with a concise, yet comprehensive, view of stem cell, molecular and cellular biology in the context of aging and age-related disease. It is hoped that this cross-disciplinary approach will foster new research that will lead to a better understanding of the role of alterations in stem cells during aging and in age-related disease, and to the development of novel approaches for manipulating stem cells in ways that promote successful aging and a longer healthspan.

CHAPTER 1

SIGNALING PATHWAYS INFLUENCING EMBRYONIC STEM CELL SELF-RENEWAL AND DIFFERENTIATION

THOMAS E. SMITHGALL

Table of contents

1.   Introduction

This chapter is focused on the signal transduction pathways that regulate stem cell self-renewal and differentiation. Murine embryonic stem (ES) cells provide a particularly useful system to address this issue. Derived from the inner cell mass of the pre-implantation mouse embryo, ES cells retain their totipotency despite continued passage in culture [1, 2]. Upon re-introduction into a developing embryo, however, ES cells can contribute to the expansion of all cell lineages. This finding has led to the

1

development of mice with defined genetic lesions, now a mainstream research tool for the understanding of gene function at the whole-animal level. Efforts are underway to develop pluripotent cell lines from other species, including humans, which may provide one avenue to the realization of stem cell therapy [3-6]. A full understanding of the molecular mechanisms that control self-renewal while maintaining pluripotency is essential to realize this goal.

2.   Maintenance of ES cell totipotency by cytokines

2.1.   Leukemia inhibitory factor (LIF) and its receptor

The first cultures of embryonic stem cells were grown on feeder layers of mouse embryo fibroblasts to suppress differentiation [1, 2]. Subsequent studies demonstrated that the feeder layer could be replaced by a single soluble cytokine, known as leukemia inhibitory factor (LIF) [7, 8]. LIF is one of several cytokines that share structural features and biological activities in a wide variety of cell types [9]. These cytokines comprise what is often referred to as the interleukin-6 (IL-6) cytokine family, which in addition to LIF and IL-6 includes oncostatin-M (OSM), cardiotropin-1 (CT-1), IL-11 and ciliary neurotrophic factor (CNTF). OSM, CNTF and CT-1 can substitute for LIF in the suppression of ES cell differentiation, suggesting some redundancy in the intracellular signaling pathways that control self-renewal [10]. IL-6 will also drive self-renewal if provided in combination with a soluble form of the IL-6 receptor [11].

The redundant biological effects of LIF and related cytokines on ES cells can be explained by the finding that they share a signal transducing receptor subunit known as gp130, a 130 kDa glycoprotein first identified as a component of the IL-6 receptor [9, 12, 13]. The receptor for LIF consists of a heterodimeric complex of gp130 and a structurally related glycoprotein known as the LIF receptor (LIFR). OSM, CT-1, and CNTF also bind to this receptor complex, although CNTF and CT-1 require a third receptor component. OSM also signals through a unique receptor component in combination with gp130, while IL-6 and IL-11 utilize two molecules of gp130 plus an additional cytokine-specific receptor subunit. As described in more detail below, studies with chimeric receptors demonstrate that the cytoplasmic region of gp130 alone is sufficient to generate signals for self-renewal. This region couples the receptor to the activation of signaling pathways critical for the maintenance of pluripotency.

2.2.   LIF activates multiple non-receptor protein-tyrosine kinases in ES cells

The first intracellular response of ES cells to LIF treatment is the induction of protein-tyrosine phosphorylation. Like other cytokine receptors, both gp130 and LIFR lack catalytic domains, and instead rely on cytoplasmic tyrosine kinases of the Jak and Src families to serve as kinase subunits. The overall structures of the cytoplasmic protein-tyrosine kinases involved in LIF signaling in ES cells are shown in Figure 1.

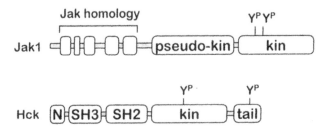

Figure 1.   Structural features of cytoplasmic tyrosine kinases implicated in LIF signaling. Assembly of LIFR/gp130 receptor complexes in response to LIF binding induces the activation of cytoplasmic protein-tyrosine kinases of the Jak and Src families in ES cells. LIF activates three members of the Jak family (Jak1, Jak2 and Tyk2) that share the structural features shown for Jak1 (*top*). These include a C-terminal catalytic domain (kin), an adjacent psuedo-kinase domain (pseudo-kin), and five additional regions of homology in the N-terminal domain that contribute to receptor interaction. Two autophosphorylation sites have been identified in the Jak1 kinase domain. Hck is the predominant member of the Src kinase family activated by LIF in ES cells (*bottom*). Hck consists of a short N-terminal unique domain (N), with sites for lipid attachment, followed by SH3, SH2 and kinase domains. The C-terminus of the protein has a short tail region with a conserved tyrosine residue, which is phosphorylated *in vivo* by the negative regulatory kinase Csk. A consensus site of autophosphorylation is localized to the kinase domain.

2.2.1. *LIF activates the Jak family of protein-tyrosine kinases in ES cells*

The Jak or Janus family of protein-tyrosine kinases currently consists of four members that are broadly involved in cytokine signaling and show the same overall structural organization (Figure 1) [9]. Three Jak kinases, Jak-1, Jak-2 and Tyk-2, are activated by LIF in ES cells. Ernst el al. [10] showed that Jak-1 and Jak-2 are activated within 2 minutes of LIF treatment, assessed as kinase autophosphorylation, and activity returned to basal levels within 60 minutes. In contrast, LIF-induced Tyk2 autophosphorylation was not observed until 20 minutes after treatment and persisted for 120 minutes. The functional significance of these kinetic differences in Jak kinase activation is not currently known.

The mechanism of kinase activation is likely to involve ligand-induced receptor oligomerization, which brings the receptor-associated kinases into close proximity and promotes their autophosphorylation by a *trans* mechanism. Detailed structural analysis of the Src tyrosine kinase family has shown that autophosphorylation within the activation loop of the catalytic domain favors an active kinase conformation [14-16]. A similar mechanism is likely to contribute to Jak activation following receptor-induced autophosphorylation; mutagenesis of tyrosine residues within the Jak2 or Tyk2 activation loop inactivates the kinase [17, 18]. Following autophosphorylation and activation, the Jak kinases phosphorylate specific tyrosine residues within the gp130 and LIFR components of the receptor complex. These tyrosine phosphorylation sites then recruit effector molecules with SH2 domains, which may become substrates for Jaks as well (see Section 3).

To assess the contribution of Jak1 to the biological effects of LIF in ES cells, Ernst et al. [10] created ES cell lines that express Jak1 antisense RNA under the control of an inducible promoter. Induction of antisense RNA suppressed Jak1 protein levels by 80%, which correlated with a 5-fold increase in the LIF concentration required to suppress differentiation. This result strongly suggests that the Jak1 signal is required for suppression of differentiation in response to LIF treatment. Interestingly, the suppression of Jak1 levels using this antisense approach did not affect LIF-induced activation of Jak2. Thus, LIF-induced Jak2 activation appears to be independent of Jak1, and suggests that the Jak2 signal cannot functionally compensate for the loss of Jak1. This finding is consistent with work using Jak1 knockout mice, in which gp130 signaling events are specifically disrupted, despite the presence of the three remaining members of the Jak kinase family. In contrast, Jak2-deficient ES cells retain their LIF responsiveness [19-21].

2.2.2.   *LIF activates the Src-related tyrosine kinase Hck in ES cells*

In addition to Jak kinases, the Src kinase family has an important role in ES cell LIF signaling as well. The Src family consists of nine members that exhibit very similar structural organization and regulation [22, 23] (Figure 1). Particularly relevant to ES cells is Hck, which exhibits an expression pattern in adults that is restricted to myeloid hematopoietic cells [24, 25]. Hck has been implicated in signal transduction pathways for a number of hematopoietic cytokines, including IL-3 and GM-CSF in addition to LIF [26-28]. Ernst et al. [10, 27] have shown that LIF induces rapid autophosphorylation of Hck in ES cells. Hck activation was first apparent 2 minutes following LIF treatment, and persisted for the duration of the assay (120 minutes). This time course strikes an interesting contrast to that of Jak1 and Jak2, which showed a transient increase that returned to basal levels within 60 minutes. One possibility is that Jak activation is a requirement for Hck activation; persistent Hck signaling may contribute to the sustained activation of the Stat3 transcription factor that is required for the maintenance of totipotency. As described in more detail in Section 3.2, Stat3 is constitutively activated in ES cells maintained in the presence of LIF [29].

Evidence suggesting that Hck activation is sufficient for self-renewal was initially reported by Ernst et al. [27]. This study demonstrated LIF-dependent association of endogenous Hck with gp130 in ES cells, consistent with a direct connection between Hck and the LIF receptor. Using gene-targeting techniques, they replaced the wild-type alleles of Hck with a constitutively active mutant form in which the negative regulatory tyrosine residue in the C-terminal tail was replaced with Phe. Cells expressing this mutant showed a dramatically reduced requirement for LIF, suggesting that Hck plays an important role in the maintenance of totipotency.

More recent work provides additional evidence that Hck is essential to gp130 signaling and ES cell self-renewal [30]. This study employed a series of chimeric receptors in which the transmembrane and cytoplasmic domains of gp130 were fused to the extracellular ligand-binding domain of G-CSF. These chimeric receptor proteins are activated by G-CSF-induced homodimerization, and generate signals indistinguishable from native gp130 in terms of self-renewal [29, 31]. By placing gp130 under the control of G-CSF, gp130 mutants can be analyzed in ES cells without interference from endogenous gp130. Using this approach, Ernst et al. [30] established a striking correlation between the ability of truncated receptors to activate Hck and to suppress ES cell differentiation. They found that mutant receptors lacking C-terminal gp130 sequences necessary for suppression of differentiation also failed to activate Hck. These truncated receptors also failed to activate the Stat3 transcription factor, which has subsequently been implicated as a key element in the self-renewal of ES cells (see Section 3.2). This result suggests that Hck may contribute to the activation of Stat3 in ES cells, a possibility that is supported by the direct activation of Stats by Hck and other Src kinases in other systems [32-35] (S. Briggs, S. Schreiner, and T. Smithgall, unpublished data). Expression of v-Src transforms ES cells to a LIF-independent phenotype [36], providing additional evidence that Src family kinases generate intracellular signals sufficient for self-renewal.

3. Signaling pathways controlling self-renewal and commitment

3.1. Effector proteins with SH2 domains bind to the tyrosine-phosphorylated LIF receptor

As described in the preceding section, LIF-induced activation of cytoplasmic tyrosine kinases leads to phosphorylation of specific tyrosine residues within the gp130 and LIFR components of the LIF receptor complex. Receptor tyrosine phosphorylation creates binding sites for signaling proteins with modular phosphotyrosine binding motifs, which include Src homology 2 (SH2) and phosphotyrosine-binding (PTB) domains [37]. This mechanism couples gp130-linked receptors to a variety of signaling pathways, including Stat transcription factors, Ras and the downstream mitogen-activated protein kinases (MAPKs) Erk1 and Erk2, and phosphatidylinositol 3-kinase. Among these, Stats (particularly Stat3) as well as the Ras/MAPK pathway have been strongly implicated in the self-renewal and differentiation responses of ES cells, respectively.

3.2. Stat transcription factors in ES cell self-renewal and differentiation

3.2.1. *Stat structure and activation*

The Stats are a family of seven structurally related transcription factors with SH2 domains that are activated in response to a wide variety of cytokines, growth factors and other polypeptide hormones [38, 39]. All Stats exhibit the same overall structure, with an N-terminal domain involved in cooperative binding to DNA, a four-helix bundle, a β-barrel which forms loops that make direct contact with DNA, an SH2 domain and tyrosine phosphorylation site which are essential for dimerization, and a C-terminal transcription transactivation domain (Figure 2).

As illustrated in Figure 2, the first step in Stat activation involves SH2-dependent recruitment to specific receptor phosphotyrosine motifs. SH2-mediated recruitment of Stat3 to the receptor places it in proximity to Jaks as well as Src kinases, both of

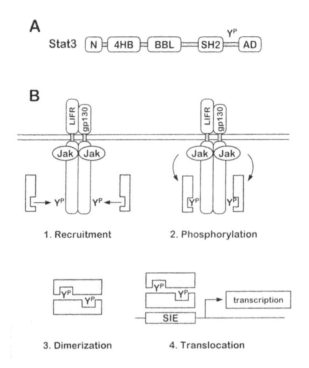

Figure 2. Mechanism of Stat3 activation by LIF. Stat3 is one of a number of transcription factors with SH2 domains that are activated by many cytokines and growth factors. A) Structurally, Stat3 consists of an N-terminal domain involved in cooperative DNA binding, a four-helix bundle (4HB), a β-barrel (BBL), an SH2 domain (SH2) and tyrosine phosphorylation site (Y^P), and a C-terminal transactivation domain (AD). B) The Stat3 activation mechanism involves several discrete steps. First, Stat3 is recruited to the activated LIFR/gp130 complex through receptor tyrosine phosphorylation sites that form a Y^PXXQ motif. Stat3 binds to this tyrosine-phosphorylated sequence through its SH2 domain. (Stat3 is represented as a shaded rectangle, with the SH2 shown as a notch.) Recruitment to the receptor places Stat3 in proximity to receptor-associated tyrosine kinases. A Jak kinase is shown, although the Src-related kinase Hck may also play a role in Stat3 phosphorylation (see text). Phosphorylation of Stat3 induces its release from the receptor and dimerization via reciprocal phosphotyrosine-SH2 interactions between the monomers. The dimeric form of Stat3 then enters the nucleus, where it interacts with promoters bearing Stat3-binding sites such as the *sis*-inducible element (SIE). LIF-induced activation of Stat3 is both necessary and sufficient for ES cell self-renewal (see text).

which may contribute to the phosphorylation of the Stat near its C-terminus [40-42]. Tyrosine phosphorylation induces Stat release from the receptor and dimerization through a reciprocal SH2-phosphotyrosine interaction. Stat dimers then enter the nucleus and bind to the consensus sequence TTCNNNGAA in the promoters of Stat-inducible genes [43]. Direct proof for the SH2 domain-dependent model of Stat dimerization is provided by the X-crystal structures of tyrosine-phosphorylated Stat dimers bound to DNA [44, 45].

3.2.2. *Stat3 activation is essential for ES cell self-renewal*

Stat3 was first described as an acute phase response factor whose DNA binding activity was induced in response to IL-6 [46, 47]. Subsequent studies have identified Stat3 as an effector protein in many cytokine and growth factor receptor systems and for v-Src and other oncogenic tyrosine kinases [38, 39, 48]. Boeuf et al. [29] investigated Stat activation in LIF-treated ES cells by electrophoretic mobility shift assay with oligonucleotide probes specific for different Stats. They observed a major LIF-dependent DNA binding activity with a probe containing a Stat3 binding site, and demonstrated a super-shift of this protein-DNA complex with a Stat3 antibody, identifying this LIF-induced DNA-binding activity as Stat3. Removal of LIF from the culture medium led to a complete loss of Stat3-DNA complexes after 12 hours. However, Stat3 protein levels were stable for 3 days after LIF withdrawal, suggesting that constitutive tyrosine phosphorylation rather than induction of the protein is responsible for its activity, most likely by Jak kinases or Hck (Section 2.2). Consistent with this view is the observation that treatment of ES cells with the tyrosine kinase inhibitor Herbimycin A blocks Stat3 DNA-binding activity and promotes differentiation despite the presence of LIF [29]. Loss of Stat3 DNA binding following LIF withdrawal could be re-induced by LIF up to 8 days later. Loss of re-induction may be due to reduced Stat3 expression in ES cells beginning 4 days after LIF withdrawal [49].

Stat3 is also controlled by phosphorylation of Ser 727 near its C-terminus [50, 51]. Mutagenesis of this serine residue decreases Stat3 transcriptional activity, suggesting that serine phosphorylation is required for full function. Recent work has implicated the Erk, p38 and Jnk MAPK pathways in Stat3 serine phosphorylation [52, 3]. These studies suggest that Stat3 may be a target for MAPK pathways activated by LIF or other factors in ES cells (see below). Consistent with this view is the finding by Boeuf et al. [29] that Stat3-DNA complexes from LIF-treated ES cells contain not only phosphotyrosine but phosphoserine as well.

The SH2 domain of Stat3 recognizes the phosphopeptide sequence Y^PXXQ, four of which have been mapped to the gp130 cytoplasmic domain and are phosphorylated in response to LIF, IL-6 and other cytokines [54]. One line of evidence implicating Stat3 as an effector for LIF in ES cells comes from truncated receptors lacking these Stat3 binding sites. Using the G-CSF receptor-gp130 chimeras described above, Ernst el al. [30] showed that deletion of all but one of these Stat3-binding motifs significantly impaired both Stat3 activation as well as self-renewal. When the tyrosine residue in the remaining Y^PXXQ motif was converted to Phe, Stat3 activation and self-renewal were completely abolished. Although this mutant receptor was unable to suppress differentiation and activate Stat3 in ES cells, it retained the capacity to induce proliferation and Erk activation in plasmacytoma cells. This finding provides evidence that distinct regions of gp130 control proliferative and differentiation responses, and are consistent with similar data in other cell types [41, 42].

In related studies, Niwa et al. [31] fused the extracellular portion of the GSF receptor with the full-length cytoplasmic region of gp130, and converted the tyrosine residues in all four of the Y^PXXQ motifs to phenylalanine. Consistent with the results from the truncated receptors described above, they found that removal of the docking sites for the Stat3 SH2 domain abolished Stat3 activation and suppression of differentiation. Studies of additional mutants revealed that the more C-terminal motifs may have a dominant role in Stat3 activation, as substitution of these sites with phenylalanine also substantially reduced self-renewal and Stat3 activation. In contrast, mutation of the membrane-proximal sites had a smaller impact on Stat3 activation and biological function in ES cells.

Matsuda et al. [55] provide similar evidence in support of the gp130-Stat3 signaling pathway in self-renewal. They created chimeric receptors between the gp130 cytoplasmic domain and the extracellular domains of the α and β chains of the GM-CSF receptor. As for the G-CSF receptor-gp130 chimeras described above, the GM-CSF receptor chimeras were fully competent to promote self-renewal following stable expression in ES cells and treatment with GM-CSF. These authors then introduced deletions and point mutations of individual gp130 tyrosine residues essential for recruitment and activation of Stat3 or SHP2, the protein-tyrosine phosphatase that couples gp130 to the activation of Ras and the Erk MAPK cascade (see Section 3.3). Using these pathway-selective receptors, they found that uncoupling gp130 from Stat3 activation prevented self-renewal. However, the receptor chimera that retained the ability to activate Stat3 but not the SHP2/Erk pathway suppressed differentiation as effectively as a chimera with the wild-type gp130 cytoplasmic domain. This result suggests Stat3 but not MAPK signaling is required for self-renewal of ES cells. However, the authors caution that the cells were passaged in the presence of fetal calf serum, which provides an alternative signal for the activation of Erk or other MAPKs. This point is important, because serine phosphorylation of Stat3 is required for full transcriptional activity [51].

Studies with dominant-negative mutants provide strong evidence that constitutive Stat3 activation is necessary for maintenance of the undifferentiated state of ES cells. Boeuf et al. [29] employed two dominant-negative mutants to investigate the contribution of Stat3 to LIF-induced transcriptional responses and self-renewal. The first mutant, Stat3F, has a phenylalanine substitution for the C-terminal tyrosine phosphorylation site. This mutant is unable to dimerize and undergo nuclear translocation, and functions in a dominant-negative manner by blocking access of endogenous Stat3 to the activated LIF receptor. The second mutant, Stat3D, is unable to bind DNA as a result of alanine substitutions for Glu 434 and Glu 435 in the DNA binding domain. Stat3D suppresses signaling by forming inactive heterodimers with endogenous Stat3. Introduction of these mutants into ES cells dramatically suppressed transcription from a LIF-dependent reporter gene construct bearing three SIE sequences tied to a minimal TK promoter [29]. (The SIE is the *sis*-inducible element derived from the c-*fos* promoter, which is commonly used as a probe for Stat3 transcriptional studies as well as gel-shift assays.) In addition, stable expression of Stat3F induced morphological differentiation of ES cells, despite their culture on feeder layers in the presence of LIF.

Niwa et al. [31] also employed Stat3F to examine the dependence of self-renewal on Stat3 activation. Their strategy involved a highly efficient episomal expression vector, obviating the need for long-term drug selection of transfected clones. Using this approach, they also observed induction of differentiation following over-expression of the Stat3F dominant-negative mutant but not wild-type Stat3. They also showed that coexpression of Stat3F with wild-type Stat3 but not Stat1 or Stat4 restored self-renewal in response to LIF treatment, providing evidence that the effects of Stat3 on differentiation are unique to Stat3 and cannot be compensated for by other Stats. Similar results were obtained in ES cells engineered to express Stat3F from an inducible promoter. In this case, induction of Stat3F expression led to ES cell differentiation and a coordinate reduction in Stat3 DNA binding activity. Taken together, these results strongly support a necessary role for Stat3 activation in the maintenance of the undifferentiated ES cell phenotype.

3.2.3. *Stat3 activation is sufficient for ES cell self-renewal.*

Although studies with LIF receptor mutants and dominant-negative Stats strongly suggest a requirement for Stat3 in ES cell self-renewal, these experiments do not address whether Stat3 activation alone is sufficient to suppress differentiation. To address this issue, Matsuda et al. [55] created a conditionally active form of Stat3 by fusing it to a modified ligand binding domain from the estrogen receptor. This modified estrogen receptor domain binds to the synthetic steroid ligand 4-hydroxytamoxifen (4HT), but not to endogenous estrogens, and has no transcriptional transactivation activity of its own [56]. The resulting fusion protein (Stat3ER) displayed strong transcriptional activity from a Stat3-inducible reporter gene construct in a 4HT-dependent manner, providing a method to selectively induce Stat3 activity independent of other pathways or transcription factors. The Stat3ER construct was then expressed in ES cells, and morphological changes were assessed in the presence of LIF, 4HT, or in the absence of factors. Cells expressing Stat3ER responded to 4HT with compact, undifferentiated colony formation that was indistinguishable from LIF-treated controls. The self-renewal capacity of the Stat3ER cells was maintained for at least 15 passages in the presence of 4HT alone. In the absence of 4HT treatment, the cells underwent differentiation in the same way as untransfected controls. To establish a specific role for Stat3 in maintaining the undifferentiated state, the authors also tested Stat5 and Stat6-estrogen receptor constructs. Neither of these constructs was able to substitute for LIF in the formation of undifferentiated colonies, supporting the idea that Stat3-dependent transcriptional events are responsible for the observed phenotype. The Stat3-ER cells were also able to contribute to chimera formation in mice, providing strong evidence that Stat3 alone is sufficient to maintain pluripotentiality. A final note concerns the phosphorylation status of the Stat3ER protein. Interestingly, treatment of the cells with 4HT induces both tyrosine and serine phosphorylation of Stat3ER, suggesting that 4HT stimulates some endogenous kinase activity in ES cells. Whether this low level of kinase activity contributes to the dimerization and activation of Stat3ER, or whether it is involved in the suppression of differentiation through other pathways requires further evaluation.

3.2.4. *Stat5 as a marker of ES cell differentiation*

Stat5a and Stat5b are two highly homologous members of the Stat family encoded by separate genes [57]. Stat5a was first described as a prolactin-inducible transcription factor in mammary gland epithelial cells [58]. Subsequent studies have revealed that both Stat5 proteins are activated by a wide variety of other cytokines, including IL-3, IL-5, GM-CSF, erythropoietin and growth hormone [59]. Stat5 activation has been associated with differentiation and survival signaling in hematopoietic cell types [41]. In particular, embryos from mice lacking both Stat5 genes are unresponsive to the anti-apoptotic effects of erythropoietin, resulting in fetal anemia [60].

Nemetz and Hocke [49] examined the role of Stat activation in ES cells and discovered an inverse correlation between Stat3 and Stat5 during differentiation. Consistent with the data summarized above, they found active Stat3 in pluripotent ES cells maintained in the presence of LIF, and a loss of Stat3 tyrosine phosphorylation following induction of differentiation by LIF withdrawal or treatment with retinoic acid. In contrast, Stat5 expression could not be detected in proliferating, pluripotent ES cells, either by immunoblotting or by RT-PCR. However, induction of differentiation led to rapid Stat5 expression, identifying Stat5 as an early marker of differentiation. Early induction of Stat5 is consistent with the requirement for this transcription factor for fetal erythropoiesis, where it couples the responses of erythoid progenitor cells to transcription of the anti-apoptotic gene, Bcl-X_L [60-63].

3.2.5. *Stats and ES cell differentiation: Future challenges*

Experiments summarized in the preceding paragraphs strongly implicate Stat3 in transcriptional events that promote ES cell self-renewal at the expense of differentiation. However, many important questions still remain. One major issue concerns the identity of the target genes for Stat3 that are ultimately responsible for these biological responses. A number of candidate genes have been identified for Stat3 in other systems, many of which are related to survival or proliferatiom (for a review, see [42]). These include the c-*myc* proto-oncogene, the mitochondrial anti-apoptotic regulators Bcl-2 and Bcl-X_L, D-type cyclins, and other proteins related to regulation of the cell cycle machinery. Whether or not these genes are regulated by Stat3 in ES cells and how they promote self-renewal while suppressing differentiation is currently unknown.

Another unresolved issue concerns the seemingly opposing actions of Stat3 in other cell types as compared to ES cells. Whereas Stat3 promotes self-renewal and suppresses differentiation of ES cells, Stat3 *induces* differentiation in other cell types. For example, in the M1 murine macrophage differentiation model, dominant-negative mutants of Stat3 block differentiation by IL-6 rather than induce it [64-66]. Other work has established that Stat3 is essential for the CNTF-dependent differentiation of astrocytes from cortical precursors [67]. One possible explanation for these differences is that Stat3 affects different transcriptional targets as a function of the maturation state or type of cell in which it is expressed. For example, Stat3 upregulates expression of the cyclin-dependent kinase inhibitor p21 in osteoblasts, an effect that correlates with gp130-mediated differentiation and suppression of apoptosis [68]. On the other hand, Stat3 promotes gp130 signals for G1 to S phase progression in pro B cells by suppressing the p21 and p27 cyclin-dependent kinase inhibitors and inducing cyclins D2, D3, and A as well as the Cdc25A phosphatase [69].

3.3.   Ras/MAPK signaling in ES cells

Another major signaling pathway regulated by LIF in ES cells is the Erk MAPK pathway. This ubiquitous pathway couples protein-tyrosine kinase signaling at the cell surface to transcriptional regulation, cytoskeletal reorganization and other responses in many systems, and contributes to cellular proliferation, differentiation and survival, depending upon the cell type. The focus of this section will be the mechanisms that relate to gp130-mediated Ras/MAPK activation and its role in ES cell differentiation. Other aspects of Ras/MAPK signaling have been reviewed extensively, and will not be considered in detail here [70-72].

3.3.1.   *The SHP2 phosphatase couples gp130 to Ras and Erk signaling*

SHP2 is a widely expressed protein-tyrosine phosphatase characterized by a C-terminal catalytic domain and tandem N-terminal SH2 domains which couple it to the activated, tyrosine-phosphorylated form of gp130 and many other receptors (Figure 3) [73]. In addition to its role as a negative regulator of protein-tyrosine kinase signaling, SHP2 serves as an essential intermediate in the activation of the Ras/MAPK pathway by gp130. A single phosphotyrosine residue in the membrane-proximal portion of the gp130 cytoplasmic domain represents the SH2 docking site for SHP2 [54]. Mutation of this tyrosine residue prevents both SHP2 interaction with gp130 and Erk activation downstream, providing evidence for an adaptor function [74]. SHP2 has two tyrosine phosphorylation sites, one of which creates a binding site for the SH2 domain of Grb2 [75, 76]. Grb2 in turn binds to the guanine nucleotide exchange factor mSos, driving Ras activation at the cytoplasmic face of the plasma membrane [77]. GTP-bound Ras recruits and activates the serine/threonine kinase Raf, leading to the sequential activation of Mek and Erk kinases downstream (Figure 3) [70]. Once activated, Erks dimerize and translocate to the nucleus, where they regulate the function of a number of transcription factors, such as the ternary complex factor Elk-1 [72, 78].

In addition to direct interaction with Grb-2/Sos, tyrosine phosphorylated SHP2 may also promote Ras activation through Gab1, an IRS-related adaptor protein [79, 80]. Gab1 has structural similarities to the *Drosophila* Dos (daughter of sevenless) protein, which is a known substrate of the *Drosophila* SHP2 homologue, Corkscrew [81, 82]. Takahashi-Tezuka et al. [79] found that Gab1 is tyrosine-phosphorylated in response to IL-6 as well as other cytokines, and tyrosine phosphorylation induced Gab1 association with SHP2 as well as PI3K. Over-expression of Gab1 enhanced gp130-mediated Erk activation, while mutation of the SHP2 docking site on gp130 disrupted Gab1 interactions with both SHP2 and PI3K as well as Erk activation. This study also showed that the the PI3K inhibitor wortmannin as well as a dominant-negative mutant of the PI3K p85 subunit blocked Erk activation, providing evidence that PI3K contributes to Erk activation downstream of gp130. Taken together, these findings suggest that Gab1 contributes to SHP2-mediated Erk activation, possibly by providing a link to the PI3K pathway. A specific role for Gab1 in ES cells has not been reported, although Gab1 is expressed and appears to be phosphorylated in response to LIF treatment in this cell type [83].

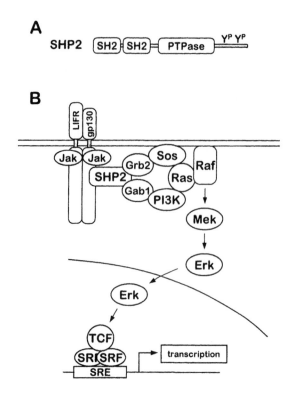

Figure 3.   SHP2 couples active LIFR/gp130 complexes to Ras/MAPK activation. A) SHP2 is a protein-tyrosine phosphatase with tandem N-terminal SH2 domains, a catalytic domain, and two C-terminal tyrosine phosphorylation sites. B) Active, tyrosine-phosphorylated LIFR/gp130 complexes recruit SHP2 through its N-terminal SH2 domains. Once recruited to the receptor, SHP2 may induce Ras/MAPK signaling through several pathways. Tyrosine phosphorylation of C-terminal tyrosines within SHP2 creates SH2 domain-binding sites for the adaptor protein Grb-2, which links SHP2 to the guanine nucleotide exchange factor, Sos. Sos activates membrane-associated Ras, which in turn activates the familiar Raf-Mek-Erk serine-threonine kinase cascade. Active Erk translocates to the nucleus where it phosphorylates ternary complex factors (TCFs). These factors bind to dimers of serum-response factor (SRF), leading to transcription from serum-response element (SRE)-linked genes. SHP2 can also associate with Gab1, an IRS-related protein. Tyrosine phosphorylated Gab1 can recruit PI3K through its p85 subunit. PI3K also interacts with Ras, and may contribute to its activation and subsequent Erk signaling downstream. SHP2 recruitment, while required for Erk activation by LIF, is not required for the maintenance of ES cell totipotency and self-renewal. However, defects in SHP2 signaling affect differentiation of multiple cellular lineages, indicating that Erk activation is essential for early stages of development (see text).

Other studies suggest that the mechanism of Ras/MAPK activation by SHP2 involves its catalytic activity in addition to its adaptor function. A catalytically inactive SHP2 mutant was shown to suppress Erk activation and Elk-1 transcriptional activity in EGF-treated 293 cells, suggesting that SHP2 phosphatase activity is required for Ras/MAPK activation in this system [84]. However, over-expression of a SHP2 mutant lacking both C-terminal tyrosine phosphorylation sites (including the Grb2 recruitment site) did not produce a dominant-negative effect, arguing against a pure adaptor function. Interestingly, the suppressive effect of the kinase-defective mutant was not observed with PDGF treatment, suggesting that SHP2 is not required for Erk activation by all growth factors.

3.3.2. *Ras/MAPK signaling contributes to differentiation commitment but is dispensable for self-renewal of ES cells*

Using the chimeric receptor approach outlined above, Burdon et al. [83] established that gp130-mediated SHP2 activation is not required for the maintenance of self-renewal. The G-CSFR/gp130 chimera used in these studies contained a mutation in the binding site for the SHP2 SH2 domain, and was therefore unable to recruit and phosphorylate SHP2 or activate Erk1 and Erk2 downstream. However, the mutant chimera retained the ability to induce the activation of Stat3, providing further evidence that the Stat3 signal alone is sufficient for the maintenance of self-renewal. Interestingly, uncoupling the receptor from SHP2 recruitment resulted in prolonged Stat3 tyrosine phosphorylation, suggesting that SHP2 serves to attenuate the Stat3 signal. A similar phenomenon has been reported in mice in which wild-type gp130 has been replaced with a mutant lacking the SHP2 recruitment site [85]. Enhanced Stat3 signaling may also account for the finding that ES cells expressing the SHP2-uncoupled chimera required 100-fold less LIF to maintain the undifferentiated phenotype [83].

Burdon et al. [83] also used a pharmacological approach to test the role of MAPK signaling in self-renewal. Using the Mek inhibitor PD098059 [86], they observed complete inhibition of LIF-induced Erk activation with 25 μM PD098059. However, this concentration of drug did not impair self-renewal. Rather, expression of a marker gene for self-renewal was found to increase in a dose-dependent manner. Importantly, ES cells propagated in the presence of PD098059 and LIF contributed to chimera formation following injection into mouse blastocysts, indicating that inhibition of gp130-induced Erk activation does not impair pluripotentiality. However, PD098059-treated ES cells cultured in the absence of LIF under conditions that promote embryoid body formation continued to express an Oct4-LacZ reporter gene. Continued activity of the Oct4 promoter is indicative of self-renewal [87-89], suggesting that lack of Erk activity interferes with differentiation. Consistent with this result is the observation that ES cells lacking serum response factor (SRF), a major transcriptional target of the Ras/MAPK pathway, fail to gastrulate and undergo mesodermal induction [90].

Genetic evidence suggesting that LIF-induced SHP2 phosphorylation and Erk activation are not necessary for self-renewal comes from studies of ES cells with mutations in the SHP2 locus. Qu et al. [91] created a line of ES cells with a homozygous deletion of the SHP2 N-terminal SH2 domain. The resulting mutant form of SHP2 retained catalytic activity, but was uncoupled from Erk activation, presumably because of a defect in SH2-mediated receptor interaction. These cells retained the capacity for self-renewal, and were found in subsequent studies to exhibit increased LIF sensitivity [92]. These results provide genetic evidence that SHP2/MAPK signaling is not required for self-renewal, and support the idea that SHP2 may serve to down-regulate Stat3 signaling and induce differentiation by coupling gp130 to the Ras/MAPK pathway.

Using the SHP2 SH2 mutant ES cells, Qu et al. [91] went on to address the SHP2 requirement for differentiation to various cellular lineages. To address the contribution of this pathway to hematopoietic differentiation, they used a two-step method in which ES cells are first permitted to form embryoid bodies (EBs). The EBs were then re-plated in colony-forming assays in the presence of cytokines that promote outgrowth of erythroid or myeloid cell types. Differentiation to both of these lineages was markedly suppressed with the SHP2 mutant ES cells. In addition, expression of the hematopoietic transcription factors GATA-1 and PU.1, which are essential for differentiation along the erythroid and myeloid pathways, respectively [93], was also markedly reduced. The authors also showed that these effects correlated with impaired MAPK activation by stem cell factor, which is required for the efficient formation of both erythroid and myeloid colonies in this assay system. These findings argue strongly for SHP2-mediated MAPK activation in the differentiation of ES cells to hematopoietic precursors.

In a subsequent study, Qu and Feng [92] demonstrated that differentiation defects in the SHP2 SH2 mutant ES cells were not limited to hematopoietic lineages. They observed that cardiomyocyte development was also substantially delayed and occurred to a lesser extent than observed with wild-type ES cells. These findings correlated with decreased expression of cardiac myosins as well as the myogenic regulatory factors myogenin and MyoD. The SHP2 mutant ES cells also failed to form epithelial cells and fibroblasts following LIF withdrawal, and instead retained a rounded, poorly differentiated morphology. These results agree with the observation that embryos derived from these SHP2 SH2 mutant ES cells die around day 9.5 of gestation and exhibit severe defects in the organization of the axial mesoderm and posterior development during gastrulation [94]. SHP2 homozygous null embryos die at the same time as the SH2 mutant, suggesting that loss of the N-terminal SH2 function and attendant uncoupling from gp130 *in vivo* is functionally the same as complete gene loss, despite the enhanced catalytic activity of the SHP2 SH2 mutant *in vitro* [95].

4.   Conclusions

Work summarized in this chapter demonstrates that distinct molecular signals control the initial developmental decisions of ES cells. LIF-induced activation of the Stat3 transcription factor through the gp130 signal transducer is both necessary and sufficient for ES cell self-renewal. Removal of LIF leads to a loss of constitutive Stat3 activation, and differentiation ensues. Initial differentiation events are dependent upon functional SHP2/Ras/MAPK signaling. Interestingly, LIF can activate both Stat3 and Erk activation simultaneously, but the Stat3 signal for self-renewal appears to be dominant. By virtue of its phosphatase activity, SHP2 may serve to provide feedback regulation of Stat3, suggesting that signals from these two pathways are integrated. Realization of the therapeutic potential of stem cells will require a more complete picture of the signaling mechanisms that control their behavior. Such information may also identify points for *ex vivo* pharmacological manipulation of stem cell self-renewal and differentiation.

5.   References

1.   Martin, G. R. (1981) Proc. Natl. Acad. Sci. USA 78, 7634-7638.

2.   Evans, M. J. and Kaufman, M. H. (1981) Nature 292, 154-156.

3.   Thomson, J. A. and Marshall, V. S. (1998) Curr. Top. Dev. Biol. 38, 133-165.

4.   Thomson, J. A., Itskovitz-Eldor, J., Shapiro, S. S., Waknitz, M. A., Swiergiel, J. J., Marshall, V. S., and Jones, J. M. (1998) Science 282, 1145-1147.

5.   Reubinoff, B. E., Pera, M. F., Fong, C. Y., Trounson, A., and Bongso, A. (2000) Nat. Biotechnol. 18, 399-404.

6.   Amit, M., Carpenter, M. K., Inokuma, M. S., Chiu, C. P., Harris, C. P., Waknitz, M. A., Itskovitz-Eldor, J., and Thomson, J. A. (2000) Dev. Biol. 227, 271-278.

7.   Williams, R. L., Hilton, D. J., Pease, S., Willson, T. A., Stewart, C. L., Gearing, D. P., Wagner, E. F., Metcalf, D., Nicola, N. A., and Gough, N. M. (1988) Nature 336, 684-687.

8.   Smith, A. G., Heath, J. K., Donaldson, D. D., Wong, G. G., Moreau, J., Stahl, M., and Rogers, D. (1988) Nature 336, 688-690.

9.   Heinrich, P. C., Behrmann, I., Muller-Newen, G., Schaper, F., and Graeve, L. (1998) Biochem. J. 334 (Pt 2), 297-314.

10.   Ernst, M., Oates, A., and Dunn, A. R. (1996) J. Biol. Chem. 271, 30136-30143.

11.   Nichols, J., Chambers, I., and Smith, A. (1994) Exp. Cell Res. 215, 237-239.

12.   Taga, T. and Kishimoto, T. (1997) Ann.Med. 29, 63-72.

13.   Hibi, M., Murakami, M., Saito, M., Hirano, T., Taga, T., and Kishimoto, T. (1990) Cell 63, 1149-1157.

14.   Sicheri, F. and Kuriyan, J. (1997) Curr. Opin. Struct. Biol. 7, 777-785.

15.   Xu, W., Doshi, A., Lei, M., Eck, M. J., and Harrison, S. C. (1999) Mol. Cell 3, 629-638.

16.   Schindler, T., Sicheri, F., Pico, A., Gazit, A., Levitzki, A., and Kuriyan, J. (1999) Mol. Cell 3, 639-648.

17.   Feng, J., Witthuhn, B. A., Matsuda, T., Kohlhuber, F., Kerr, I. M., and Ihle, J. N. (1997) Mol. Cell Biol. 17, 2497-2501.

18.   Gauzzi, M. C., Velazquez, L., McKendry, R., Mogensen, K. E., Fellous, M., and Pellegrini, S. (1996) J. Biol. Chem. 271, 20494-20500.

19.   Parganas, E., Wang, D., Stravopodis, D., Topham, D. J., Marine, J. C., Teglund, S., Vanin, E. F., Bodner, S., Colamonici, O. R., van Deursen, J. M., Grosveld, G., and Ihle, J. N. (1998) Cell 93, 385-395.

20.   Neubauer, H., Cumano, A., Muller, M., Wu, H., Huffstadt, U., and Pfeffer, K. (1998) Cell 93, 397-409.

21.   Rodig, S. J., Meraz, M. A., White, J. M., Lampe, P. A., Riley, J. K., Arthur, C. D., King, K. L., Sheehan, K. C., Yin, L., Pennica, D., Johnson, E. M., Jr., and Schreiber, R. D. (1998) Cell 93, 373-383.

22.   Corey, S. J. and Anderson, S. M. (1999) Blood 93, 1-14.

23.   Lowell, C. A. and Soriano, P. (1996) Genes Dev. 10, 1845-1857.

24.   Ziegler, S. F., Marth, J. D., Lewis, D. B., and Perlmutter, R. M. (1987) Mol. Cell. Biol. 7, 2276-2285.

25.   Quintrell, N., Lebo, R., Varmus, H., Bishop, J. M., Pettenati, M. J., Le Beau, M. M., Diaz, M. O., and Rowley, J. D. (1987) Mol. Cell. Biol. 7, 2267-2275.

26.   Linnekin, D., Howard, O. M. Z., Park, L., Farrar, W., Ferris, D., and Longo, D. L. (1994) Blood 84, 94-103.

27.   Ernst, M., Gearing, D. P., and Dunn, A. R. (1994) EMBO J. 13, 1574-1584.

28. Anderson, S. M. and Jorgensen, B. (1995) J. Immunol. 155, 1660-1670.

29. Boeuf, H., Hauss, C., Graeve, F. D., Baran, N., and Kedinger, C. (1997) J. Cell Biol. 138, 1207-1217.

30. Ernst, M., Novak, U., Nicholson, S. E., Layton, J. E., and Dunn, A. R. (1999) J. Biol. Chem. 274, 9729-9737.

31. Niwa, H., Burdon, T., Chambers, I., and Smith, A. (1998) Genes Dev. 12, 2048-2060.

32. Garcia, R., Yu, C.-L., Hudnall, A., Catlett, R., Nelson, K., Smithgall, T. E., Fujita, D. J., Ethier, S., and Jove, R. (1998) Cell Growth and Differentiation 8, 1267-1276.

33. Turkson, J., Bowman, T. L., Garcia, R., Caldenhoven, E., de Groot, R. P., and Jove, R. (1998) Mol. Cell. Biol. 18, 2545-2552.

34. Yu, C.-L., Meyer, D. J., Campbell, G. S., Larner, A. C., Carter-Su, C., Schwartz, J., and Jove, R. (1995) Science 269, 81-83.

35. Cao, X. M., Tay, A., Guy, G. R., and Tan, Y. H. (1996) Mol. Cell.Biol. 16, 1595-1603.

36. Boulter, C. A., Aguzzi, A., Williams, R. L., Wagner, E. F., Evans, M. J., and Beddington, R. (1991) Development 111, 357-366.

37. Kuriyan, J. and Cowburn, D. (1997) Annu. Rev. Biophys. Biomol. Struct. 26, 259-288.

38. Ihle, J. N. (1996) Cell 84, 331-334.

39. Darnell Jr., J. E. (1997) Science 277, 1630-1635.

40. Reddy, E. P., Korapati, A., Chaturvedi, P., and Rane, S. (2000) Oncogene 19, 2532-2547.

41. Smithgall, T. E., Briggs, S. D., Schreiner, S., Lerner, E. C., Cheng, H., and Wilson, M. B. (2000) Oncogene 19, 2612-2618.

42. Hirano, T., Ishihara, K., and Hibi, M. (2000) Oncogene 19, 2548-2556.

43. Hoey, T. and Schindler, U. (1998) Curr. Opin. Genet. Dev. 8, 582-587.

44. Chen, X., Vinkemeier, U., Zhao, Y., Jeruzalmi, D., Darnell, J. E., Jr., and Kuriyan, J. (1998) Cell 93, 827-839.

45. Becker, S., Groner, B., and Muller, C. W. (1998) Nature 394, 145-151.

46. Zhong, Z., Wen, Z., and Darnell, J. E., Jr. (1994) Science 264, 95-98.

47. Akira, S., Nishio, Y., Inoue, M., Wang, X. J., Wei, S., Matsusaka, T., Yoshida, K., Sudo, T., Naruto, M., and Kishimoto, T. (1994) Cell 77, 63-71.

48. Bowman, T., Garcia, R., Turkson, J., and Jove, R. (2000) Oncogene 19, 2474-2488.

49. Nemetz, C. and Hocke, G. M. (1998) Differentiation 62, 213-220.

50. Wen, Z. and Darnell, J. E., Jr. (1997) Nucleic Acids Res. 25, 2062-2067.

51. Wen, Z., Zhong, Z., and Darnell, J. E., Jr. (1995) Cell 82, 241-250.

52. Turkson, J., Bowman, T., Adnane, J., Zhang, Y., Djeu, J. Y., Sekharam, M., Frank, D. A., Holzman, L. B., Wu, J., Sebti, S., and Jove, R. (1999) Mol. Cell Biol. 19, 7519-7528.

53. Chung, J., Uchida, E., Grammer, T. C., and Blenis, J. (1997) Mol. Cell Biol. 17, 6508-6516.

54. Stahl, N., Farruggella, T. J., Boulton, T. G., Zhong, Z., Darnell Jr., J. E., and Yancopoulos, G. D. (1995) Science 267, 1349-1353.

55. Matsuda, T., Nakamura, T., Nakao, K., Arai, T., Katsuki, M., Heike, T., and Yokota, T. (1999) EMBO J. 18, 4261-4269.

56. Littlewood, T. D., Hancock, D. C., Danielian, P. S., Parker, M. G., and Evan, G. I. (1995) Nucleic Acids Res. 23, 1686-1690.

57. Liu, X., Robinson, G. W., Gouilleux, F., Groner, B., and Hennighausen, L. (1995) Proc. Natl. Acad. Sci. USA 92, 8831-8835.

58. Wakao, H., Gouilleux, F., and Groner, B. (1994) EMBO J. 13, 2182-2191.

59. Ihle, J. N., Stravapodis, D., Parganas, E., Thierfelder, W., Feng, J., Wang, D., and Teglund, S. (1998)

Cancer J. Sci. Am. 4 Suppl 1, S84-S91.

60. Socolovsky, M., Fallon, A. E., Wang, S., Brugnara, C., and Lodish, H. F. (1999) Cell 98, 181-191.

61. Silva, M., Benito, A., Sanz, C., Prosper, F., Ekhterae, D., Nunez, G., and Fernandez-Luna, J. L. (1999) J. Biol. Chem. 274, 22165-22169.

62. Silva, M., Grillot, D., Benito, A., Richard, C., Nunez, G., and Fernandez-Luna, J. L. (1996) Blood 88, 1576-1582.

63. Gregoli, P. A. and Bondurant, M. C. (1997) Blood 90, 630-640.

64. Nakajima, K., Yamanaka, Y., Nakae, K., Kojima, H., Ichiba, M., Kiuchi, N., Kitaoka, T., Fukada, T., Hibi, M., and Hirano, T. (1996) EMBO J. 15, 3651-3658.

65. Yamanaka, Y., Nakajima, K., Fukada, T., Hibi, M., and Hirano, T. (1996) EMBO J. 15, 1557-1565.

66. Minami, M., Inoue, M., Wei, S., Takeda, K., Matsumoto, M., Kishimoto, T., and Akira, S. (1996) Proc. Natl. Acad. Sci. USA 93, 3963-3966.

67. Bonni, A., Sun, Y., Nadal-Vicens, M., Bhatt, A., Frank, D. A., Rozovsky, I., Stahl, N., Yancopoulos, G. D., and Greenberg, M. E. (1997) Science 278, 477-483.

68. Bellido, T., O'Brien, C. A., Roberson, P. K., and Manolagas, S. C. (1998) J. Biol. Chem. 273, 21137-21144.

69. Fukada, T., Ohtani, T., Yoshida, Y., Shirogane, T., Nishida, K., Nakajima, K., Hibi, M., and Hirano, T. (1998) EMBO J. 17, 6670-6677.

70. Campbell, S. L., Khosravi-Far, R., Rossman, K. L., Clark, G. J., and Der, C. J. (1998) Oncogene 17, 1395-1413.

71. Robinson, M. J. and Cobb, M. H. (1997) Curr. Opin. Cell Biol. 9, 180-186.

72. Cobb, M. H. and Goldsmith, E. J. (2000) Trends Biochem. Sci. 25, 7-9.

73. Neel, B. G. and Tonks, N. K. (1997) Curr. Opin. Cell Biol. 9, 193-204.

74. Fukada, T., Hibi, M., Yamanaka, Y., Takahashi-Tezuka, M., Fujitani, Y., Yamaguchi, T., Nakajima, K., and Hirano, T. (1996) Immunity 5, 449-460.

75. Bennett, A. M., Tang, T. L., Sugimoto, S., Walsh, C. T., and Neel, B. G. (1994) Proc. Natl. Acad. Sci. USA 91, 7335-7339.

76. Li, W., Nishimura, R., Kashishian, A., Batzer, A. G., Kim, W. J., Cooper, J. A., and Schlessinger, J. (1994) Mol. Cell Biol. 14, 509-517.

77. Schlessinger, J. (1993) Trends Biochem. Sci. 18, 273-275.

78. Wasylyk, B., Hagman, J., and Gutierrez-Hartmann, A. (1998) Trends Biochem. Sci. 23, 213-216.

79. Takahashi-Tezuka, M., Yoshida, Y., Fukada, T., Ohtani, T., Yamanaka, Y., Nishida, K., Nakajima, K., Hibi, M., and Hirano, T. (1998) Mol. Cell Biol. 18, 4109-4117.

80. Holgado-Madruga, M., Emlet, D. R., Moscatello, D. K., Godwin, A. K., and Wong, A. J. (1996) Nature 379, 560-564.

81. Perkins, L. A., Larsen, I., and Perrimon, N. (1992) Cell 70, 225-236.

82. Herbst, R., Carroll, P. M., Allard, J. D., Schilling, J., Raabe, T., and Simon, M. A. (1996) Cell 85, 899-909.

83. Burdon, T., Stracey, C., Chambers, I., Nichols, J., and Smith, A. (1999) Dev. Biol. 210, 30-43.

84. Bennett, A. M., Hausdorff, S. F., O'Reilly, A. M., Freeman, R. M., and Neel, B. G. (1996) Mol. Cell Biol. 16, 1189-1202.

85. Ohtani, T., Ishihara, K., Atsumi, T., Nishida, K., Kaneko, Y., Miyata, T., Itoh, S., Narimatsu, M., Maeda, H., Fukada, T., Itoh, M., Okano, H., Hibi, M., and Hirano, T. (2000) Immunity 12, 95-105.

86. Dudley, D. T., Pang, L., Decker, S. J., Bridges, A. J., and Saltiel, A. R. (1995) Proc. Natl. Acad. Sci. USA 92, 7686-7689.

87. Dani, C., Chambers, I., Johnstone, S., Robertson, M., Ebrahimi, B., Saito, M., Taga, T., Li, M., Burdon, T., Nichols, J., and Smith, A. (1998) Dev. Biol. 203, 149-162.

88. Rosner, M. H., Vigano, M. A., Ozato, K., Timmons, P. M., Poirier, F., Rigby, P. W., and Staudt, L. M. (1990) Nature 345, 686-692.

89. Okamoto, K., Okazawa, H., Okuda, A., Sakai, M., Muramatsu, M., and Hamada, H. (1990) Cell 60, 461-472.

90. Arsenian, S., Weinhold, B., Oelgeschlager, M., Ruther, U., and Nordheim, A. (1998) EMBO J. 17, 6289-6299.

91. Qu, C. K., Shi, Z. Q., Shen, R., Tsai, F. Y., Orkin, S. H., and Feng, G. S. (1997) Mol. Cell Biol. 17, 5499-5507.

92. Qu, C. K. and Feng, G. S. (1998) Oncogene 17, 433-439.

93. Orkin, S. H. (2000) Nature Reviews Genetics 1, 57-64.

94. Saxton, T. M., Henkemeyer, M., Gasca, S., Shen, R., Rossi, D. J., Shalaby, F., Feng, G. S., and Pawson, T. (1997) EMBO J. 16, 2352-2364.

95. Arrandale, J. M., Gore-Willse, A., Rocks, S., Ren, J. M., Zhu, J., Davis, A., Livingston, J. N., and Rabin, D. U. (1996) J. Biol. Chem. 271, 21353-21358.

CHAPTER 2

HEMATOPOIETIC STEM CELLS AND AGING

GARY VAN ZANT, ERIN L. MANNING and HARTMUT GEIGER

Table of contents

Stem Cells: A Cellular Fountain of Youth. Ed. by Mark P. Mattson and Gary Van Zant. 19 — 42
© 2002 *Elsevier Science B.V. All rights reserved.*

1. Introduction

Self-renewing tissues rely on stem cell populations to continuously supply those mature cells with inherently short lifespans or those lost through wear and tear, disease, or accident. Stem cells must therefore function throughout the lifespan of an animal; failure to do so would result in life-threatening aplasias. Since stem cells are capable of not only supplying differentiated cells to a tissue, but also replicating to resupply their own numbers, it would seem that such populations would be exempt from age-related depletion. An alternative perspective is that they may be exquisitely sensitive to age-related functional decline if their regenerative properties are not fully maintained. Using lympho-hematopoiesis as a model, it is widely recognized that production of most blood cells is maintained at near-normal levels throughout the lifespans of rodents and humans. An exception is the redistribution in lymphoid cell subpopulations during aging resulting in the contraction of some subsets, and, of course, the involution of the thymus. It is further regognized that the response to hematological and immunological stress is blunted in older animals and humans despite the fact that in mice the number of stem cells actually increases in the bone marrow during normal aging. However, a growing body of evidence shows that despite an age-related increase in at least hematopoietic stem cell numbers, some functional capabilities of stem cells are compromised during aging. We discuss our recent studies using a genetic approach in mice to identify loci regulating stem cell populations during the aging process. Our data point to the notion that an age-related decline in stem cell function, perhaps due to replicative stress, directly affects organismal longevity.

2. Stem cell populations drive developmental systems

2.1. Models of stem cell differentiation

In adult, self-renewing tissues, the stem cell population serves as a continuous source of the appropriate differentiated cells throughout the lifespan of the animal. Thus, over an extended period, this small cell population, acting in response to a complex series of regulatory controls both cell-intrinsic and cell-extrinsic, must parse differentiation and replication in such a way as to maintain its numbers while at the same time providing enough mature cells for proper physiological function. It is not the aim here to provide an exhaustive review of stem cell models; for this the reader is directed to a volume of reviews in which these issues are more fully developed [1]. However, in brief, a stem cell population may meet these demands through two broad mechanisms:

2.1.1. *Clonal succession*

One possible mechanism specifies that the majority of stem cells are maintained in a quiescent, noncycling state until needed and then one (or a few) at a time is recruited and activated to become the active clone(s) [2]. Such a model preserves primitive developmental potential in the founder stem cell population through dormancy, but without additional features does not allow for expansion of the stem cell pool through self-renewal. It also implies that, despite extensive developmental potential of individual stem cells, clonal longevity is limited and thus individual stem cell clones have a finite lifespan.

2.1.2. *Alternate fates of daughter cells produced by stem cell division*

A second possible mechanism relies on stem cell proliferation to provide flexibility and preserve a stem cell reserve. At each such cell division, depending upon demands, the stem cell, through symmetric or asymmetric divisions, may give rise to (a) new stem cells, (b) progenitor cells one step removed from stem cells but potentially still pluripotent, or (c) a combination of the two cell types. Despite the fact that Osgood proposed such a model of alpha cell division in the 1950s [3], little is known of how the regulatory controls might work. Extrinsic and/or instrinsic signaling mechanisms, if any, specifying which type of cell division a stem cell is to carry out are presently obscure. It could be a stochastic event or it could be deterministic in nature through responses to extrinsic signals derived from the local milieu (cytokines and stroma). Alternatively, deterministic signals could be generated intrinsically by the stem cell itself; for example by partitioning a cell component at mitosis. Such a mechanism is employed at the earliest cell divisions of embryogenesis when preformed cell components of the zygote are partitioned to specify the earliest restrictions in cell fate among blastomeres [4].

2.1.3. *Stochastic mechanisms in stem cell function*

A stem cell population *in toto* may efficiently orchestrate an appropriate response to specific hematopoietic needs, whereas at the level of individual stem cells all mitotic options, symmetric and asymmetric, may be open to each cell but with variable probabilities that reflect the current needs [5]. Thus, events at the individual cell level may be stochastic, but the population is responsive to physiological needs [6]. It should be pointed out that at least the major features of the two models do not necessarily need to be mutually exclusive. For instance, the active stem cell clone in a succession model may have the options open to it present in a stochastic process, provided that stem cell division may produce other stem cells. Irrespective of the single or multiple mechanism(s) of stem cell control, we know from a large body of experimental and clinical data that stem cell--driven developmental systems have considerable inherent flexibility. In fact, it appears to be sufficiently robust that maintenance of the stem cell population should be assured under a variety of experimentally created and naturally occurring conditions including aging.

2.2. A scarcity of stem cells complicates their study

Because of their numerical paucity in tissues and their lack of distinguishing morphological characteristics, stem cells have been, and continue to be, difficult to study. As has been previously noted, because of these mitigating properties, stem cells have been studied by their functional abilities -- proliferative and developmental potencies.

Historically stemming from concerns about the health effects of radiation following the advent of nuclear energy and the use of nuclear weapons, the hematopoietic stem cell has been the focus of concerted studies over many years and has arguably yielded the largest body of information about stem cells in general. However, as discussed in other chapters in this volume, revelations of extensive flexibility in developmental potential make it increasingly likely that findings obtained with one stem cell will be generally applicable to others.

Despite considerable efforts, it has only been in the last 10 years or so that stem cell purification has been widely available to workers in the field, permitting their detailed prospective study. Expression of one set of cell surface antigens, and the lack of expression of other differentiation markers, was correlated with functional stem cell activity such that it became possible to use fluorochrome-tagged antibodies and fluorescence-activated cell sorting and/or immuno-absorptive methods to isolate viable stem cells. Due to the pioneering developmental efforts of several labs, it is now reasonably straightforward to obtain purified stem cell populations [7-9]. Purified stem cell populations contain virtually all of the long-term repopulating ability of bone marrow as measured by transplantation into lethally irradiated mice [10], or, in the case of human stem cells, to durably engraft immunodeficient mice bearing mutations for nonobese diabetes and severe combined immunodeficiency (NOD-SCID mice) [11]. Similar advances have recently been made in the purification of neural stem cells, drawing upon the techniques used for hematopoietic stem cell isolation [12, 13].

2.3.   Stem cell renewal after transplantation

A widely accepted assumption held by most stem cell biologists is that stem cells are generally capable of self-renewal such that daughter stem cells (but not derived progenitor cells) are functionally equivalent to those of the previous generation, and thus that stem cells are an inexhaustible cellular source. There is good reason for this and it comes in part from experience with both mice and humans in the field of bone marrow transplantation. In this procedure, a minute fraction of the original stem cell pool is transplanted and is required to satisfy not only the immediate need for mature blood cells in the ablated recipient, but also to rebuild the coordinated hierarchy of stem, progenitor, and maturing cells necessary to maintain lympho-hematopoiesis over the long term. That such a small stem cell graft can accomplish this daunting task appears to beg the question as to whether stem cell renewal occurs. Limiting dilution transplant experiments using highly purified stem cells have convincingly shown that even a single cell is capable of this dramatic feat [14, 15]. Moreover, serial transplant studies in mice have shown that the original graft may not only repopulate the primary host but if marrow from these recipients is subsequently harvested and transplanted into and passed through secondary, tertiary, and quaternary hosts, the original, small stem cell population can cumulatively provide differentiated cells for longer than the lifespan of the original donor animal [16, 17].

Consequently, there is little question that a functional stem cell population can increase in size and therefore that stem cells can replicate to produce other cells with the cardinal functional characteristics of stem cells. Considerable light has been shed on this issue by the use of stem cells that have been genetically marked by proviral insertion. Since the genomic insertion pattern for each stem cell and its progeny is unique, stem cell pedigrees can be established. Several groups have shown that retrovirally marked stem cells could replicate either *in vitro* or *in vivo* and that their progeny retained developmental potency to subsequently engraft transplant recipients long-term [18-20]. However, there is also abundant data showing that the stem cell population is qualitatively and quantitatively diminished after transplant [14]. The fact that one cannot serially transplant stem cells indefinitely is but one finding supporting this. Similarly, abundant clinical data have shown that patients who have undergone extensive chemotherapy, not only possess fewer cells capable of autologous engraftment, but the tempo of engraftment, when it does occur, is slow. Long-term repopulating stem cells recover after transplant in the mouse to only about 4% of normal values in 6 weeks, and do not further increase even after many months [21]. Despite the blunted recovery, absolute numbers do increase by about ten-fold over the numbers in the graft used for each of the transplants. The fact that the ten-fold expansion is found irrespective of the number of stem cells in the graft, led Iscove and Nawa to propose that extrinsic factors in the recipient, such as cytokines, limited the level of expansion rather than replicative stress, exhaustion, and senescence.

Retrovirally marked stem cells have also been used to quantify contributions of individual stem cell clones to engraftment after transplant. In aggregate, the results have shown that short-term engraftment was accomplished by many individual clones but that after a few months an oligoclonal pattern emerged [22]. Some, but not all, clones persisted for periods of time that approximated the mouse lifespan [23-25], and in extreme examples a single clone maintained all hematopoiesis in a transplant recipient for a year or more [26].

2.4.   Engraftment after stem cell transplant is not homeostatic

It is clear that fundamental differences exist between steady-state hematopoiesis and engraftment after stem cell transplant which may affect stem cell regulation. The disparity in the sizes of the natural stem cell pool compared with the small number in a graft has already been alluded to. Moreover, it has been argued that the experimental manipulations involved in carrying out a stem cell transplant, with respect to both the donor and the recipient, are sufficient to make extrapolation to natural controls invalid [27]. The disruption of the stem cells from their close association with stromal cells comprising their local microenvironment, their injection into the venous circulation of the recipient, and the perhaps selective requirement that stem cells lodge in an appropriate new micorenvironmental niche, introduces large variation from steady-state conditions. Similarly for the recipient, treatment with large doses of ionizing radiation and/or chemotherapeutic drugs has significant but not fully defined effects on the stromal cells and the resulting cytokine titers to which the stem cells are exposed.

In contrast to results obtained from serial transplantation, others have shown that a protocol involving 25 cycles of hydroxyurea injections over a year did not measurably diminish stem cell function, despite the fact that the cumulative replicative requirement was quantitatively similar to serial transplantation [28]. These data suggest an inherent difference between experimental measurements of stem cell function made using transplantation models and those made *in vivo* without transplantation.

Consistent with these results were findings obtained with chimeric mice constructed by combining two ES cell lines bearing distinguishable markers expressed in mature blood cells, but otherwise sharing the same genetic background [29]. They were characterized by stable, chimeric blood cell populations over long periods of time, suggesting the simultaneous contributions of a large number of stem cells of both genotypes rather than oligoclonal participation and clonal succession. Study of chimeras constructed by aggregating embryos of different mouse strains (allophenic mice) have provided data consistent with both models [30, 31]. Some chimeras, at certain times in their life histories, exhibited dramatic fluctuations in chimerism of blood cells, consistent with a clonal succession model, whereas the majority displayed chimerism consistent with the simultaneous participation of a large number of stem cell clones. As discussed below, the use of allophenic mice has provided insight into competitive advantages of genotype-restricted stem cell clones, since in this model embryos of different genetic backgrounds were combined, and has led to the identification of several stem cell genes affecting their cell cycle kinetics and population sizes in the bone marrow.

2.5.　Stem cells cycle slowly but continuously

In persuasive studies aimed at resolving the issue of stem cell cycling, Bradford et al. [32] and Cheshier et al. [33] have confirmed and extended original kinetic studies of Pietrzyk et al. [34], by labeling hematopoietic stem cell populations of normal, unmanipulated mice through the long-term addition of bromo-deoxyuridine (BrdUrd) to their drinking water. The results have shown that essentially the entire stem cell population is slowly labeled over 2-3 months in a pattern consistent with virtually all cells having passed through at least one cell cycle during this period. Following engraftment of limiting dilutions of genetically marked stem cells, another group used stochastic modeling to estimate that stem cells replicate somewhat more frequently (on average every 2-3 weeks) in a transplantation model [35]. Given the fact that only a small percentage of stem cells are in cycle at any given time, the data suggest that stem cells may repetitively enter and leave cell cycle from a quiescent G_o state. In contrast, rapidly dividing mouse cells usually have a 12-24 h cell cycle time, with essentially none being out of cell cycle in quiescent G_o. These data argue strongly against a succession of stem cell clones, since such a mechanism would show labeling in only a small fraction of stem cells during the period of exposure to BrdUrd — in theory, perhaps only one. The results also corroborate the findings in chimeras showing stable blood cell chimerism over long time periods and further underscore the fundamental difference between stem cell regulation in unmanipulated animals and in a transplant model.

2.6. Are mouse studies representative?

Much of the experimental evidence on stem cell usage comes from studies of mice, but there is some evidence in larger mammals consistent with an oligoclonal, active stem cell population. For example, autologous transplant studies in cats have shown that allelic expression of an X-linked marker in marrow cells of heterozygous females fluctuates for several years after transplant, but with time becomes stable and mature blood cell populations may express only one or the other G6PD alleles [36]. These results, like the pattern of clonal expansion of genetically marked and transplanted stem cells in mice, are consistent with few stem cells clones being simultaneously active both short- and long-term after transplant. Such an experimental approach took advantage of the fact that only one X chromosome is transcriptionally active in adult, female cells and that X-inactivation is random in embryonic cells, including the founder hematopoietic stem cell population. Similarly, in older women heterozygous for allelic variants of G6PD, there is increased and apparently random skewing toward one variant or the other during aging [37]. Such data are consistent with increasingly fewer clones contributing to hematopoiesis in old age, and provide additional evidence suggesting a small, active stem cell population.

Further evidence supporting a difference between stem cells of mice and larger mammals has to do with their frequency and rate of turnover in the bone marrow. Estimates of stem cell frequency and replication from stochastic modeling of limiting dilution transplants in mice and cats, have shown murine stem cells occur at nearly a hundred-fold higher frequency than in the cat (8 per 10^5 vs. 6 per 10^7) and replicate on average every 2.5 weeks rather than once in 8-10 weeks as in the cat [35]. As these authors speculate, there may be a relationship between stem cell numbers, periodicity of stem cell replication and lifespan. Cats have a maximal lifespan of roughly 15 years whereas in mice 2-3 years is usually maximum. In following sections we discuss our studies aimed at identifying genetic determinants of natural variation in stem cell population size, replication rate and longevity in mice.

Despite obvious differences in size and longevity in mammalian species, it seems unlikely that stem cell population organization and usage would be fundamentally different between mice and larger mammals including humans. Rather, given the different methods of study, technical details of experimental procedures and assays may account for the differences. For example, in the transplantation studies in cats, stem cells were transplanted at limiting dilutions such that some recipients received too few stem cells to engraft necessitating a second transplant with more stem cells. It has been shown previously that the number of stem cells transplanted affects not only the clonal distribution of cells amongst the host's hematopoietic tissues, but also the contributions from individual cells during engraftment [38]. As one might expect, the greater the number of stem cells transplanted, the larger the number contributing to engraftment. At limiting dilutions, oligoclonal hematopoiesis might be expected to be the norm as illustrated experimentally in the study by Micklem [38].

The issue of X chromosome inactivation and somatic cell clonality in elderly women may be complicated by additional factors. Recent studies have shown that the X chromosome may contain genes relevant to allele-specific skewing that are deterministic rather than representing stochastic events or a diminution in the number of active clones over time. In accordance with other studies, Buller et al. [39] found nonrandom X chromosome inactivation patterns in neoplastic tissue that have usually been interpreted as evidence of the clonal origin of tumors. However, they hypothesize that the data in this case are consistent with the existence of an X-linked tumor suppressor gene. A germline mutation of this gene combined with nonrandom X-inactivation may reduce or abolish expression of the wild-type allele and predispose these women to ovarian and/or breast cancer. Underscoring the importance of deterministic X-linked alleles, it has been shown in certain heterozygous cats that clonal dominance during aging is due to a competitive advantage conferred by a locus on the X chromosome that possibly acts by enhancing proliferation or decreasing cell death in the affected clone [40].

2.7.   Apoptosis in continuously renewing stem cell systems

The foregoing discussion has centered on the issues of stem cell division and the developmental cell fate of stem cell progeny. The role of apoptosis has been neglected although it may play an important role in determining population dynamics of stem, progenitor and maturing cell populations, as it does during development [41]. Dysregulation of tumor-suppressing genes in cancer, allowing tumor cells to bypass normal population growth restraints through apoptosis, has emphasized the importance of cell death in the process of population control. Conditional survival of cells at all levels in the developmental hierarchy, including stem cells, may involve a variety of cell-extrinsic and cell-intrinsic factors preventing or enabling apoptosis. Transgenic mice over-expressing BCL-2, an anti-apoptotic tumor-suppressor, show about a two-fold increase in stem cell numbers, suggesting that apoptosis is a natural, but modest, regulator of stem cells *in vivo* [42]. Moreover, in competitive repopulation experiments of lethally irradiated recipient mice, marrow from Bcl-2 transgenic mice had a distinct competitive repopulation advantage, suggesting that even during high-demand hematopoiesis accompanying engraftment, apoptosis plays a role in diminishing potential stem and progenitor cell differentiation and output [43]. In support of these results, stochastic modeling of stem cell engraftment in both mice and cats has shown that apoptosis of stem cells is a small, but estimable factor in the dispensation of stem cells [35].

3.   Hematopoietic stem cells as a model population for studies of aging

3.1.   Access and means of study

Hematopoietic stem cells have been the subject of extensive studies, including effects of aging, for several reasons; (a) their mature progeny, blood cells, are readily accessible in the peripheral blood and this source can be sampled repetitively from the same animal or patient over long time periods; (b) stem cell transplantation is used extensively in the clinical treatment of cancer, heightening research in the area; and (c) the hematopoietic tissues containing stem cells, principally bone marrow in adults, are reasonably accessible.

The diffuse nature of hematopoietic tissue has been both a curse and a blessing to its study. Diffuse organization has made it difficult to establish the histological organization of the marrow, particularly with respect to the interactions of hematopoietic cells and supporting stroma and vasculature. At the other end of the spectrum is the discrete organization of the crypts and villi of the intestinal epithelium where the locations and numbers of stem cells, progenitors, and maturing cells follow precisely defined and discernible patterns [44, 45]. On the other hand, hematopoietic tissues are easily handled and, most importantly, it is very easy to obtain viable single cell suspensions. The latter point has allowed the field to benefit from flow cytometry, a powerful analytical and cell separation tool. Combined with advances in cell culture, these have led to an increasingly detailed, but still incomplete, developmental map of the system.

3.2.   Do stem cells age?

The path to a clear view of the effects of aging on the hematopoietic stem cell population has been, and remains, difficult [for additional perspectives, see Morrison et al. [46] and Globerson [47]]. Whether or not stem cell populations age is an important question, with significant clinical ramifications, to which answers are still being sought. There is a wealth of data on this issue, with some of it being contradictory. On one hand, evidence points to an essentially unlimited renewal capacity, while on the other there are clear indications of diminished functional responses to replicative stress. The contradictions arise, at least in part, because of the myriad technical approaches taken and assays used in this pursuit. Diversity rather than a more standardized approach is rooted in the scarcity of stem cells and, at least historically, their lack of easily distinguished characteristics. More rapid progress has come since fairly well standardized cell separation protocols have been available for the purification of mouse and human stem cells. As previously discussed, using flow cytometry and sorting, it is now possible to obtain fairly large populations of highly purified stem cells for research and even clinical transplantation.

Although the size and composition of mature blood cell populations remains largely unchanged into old age, functional attributes of some mature cells are compromised [48]. For example, biochemical changes in red cell membranes shorten their lifespan in the circulation, which may alter their oxygen exchange rates and capacities [49]. In addition, the distribution of lymphocyte subpopulations is altered, the cytokine profiles they produce change, and at least some aspects of surveillance and disease fighting are diminished [50, 51].

Despite their ability to continuously supply mature cells at normal levels throughout life and their ability under extreme experimental conditions, such as bone marrow transplantation, to meet demands not normally encountered, the literature generally, but not universally, supports the contention that stem cells age. For example, early studies by Ogden and Micklem [52] determined the rate of decline in stem cells during serial transplantation of marrow from young and old donors. They found that stem cells in old marrow were capable of fewer repetitive transfers than those in young marrow. Similarly, Micklem et al. [53] found a preferential engraftment advantage using fetal liver rather than adult mouse marrow to transplant lethally irradiated recipients. In concurrence, Mauch et al. [54] found similar decrements in the proliferative capacity of marrow from old animals when transplanted.

However, other studies of the functional properties of bone marrow from aged mice, showed that old stem cells engrafted irradiated recipients at least equally as well as, and in some cases more effectively than, marrow from young adult mice [55]. Central to these and to later studies helping to clarify the issue, Harrison developed a competitive repopulation assay in which mixtures of young and old marrow or fetal liver were used to uncover potential functional differences in cell populations to engraft and maintain hematopoiesis long-term [56]. Distinguishable genetic markers between the cell sources enabled one to quantify and determine the genetic derivations of hematopoietic cells.

The original study showing no diminished capacity of old stem cells was carried out by mixing specified numbers of marrow cells from young and old marrow to comprise the graft. Several groups have subsequently shown that the number of stem cells, identified either by cell surface phenotype or functionally, in most mouse strains actually increased with age [46, 57, 58]. Thus, at least part of the repopulation results could be explained by the fact that the frequency of stem cells was higher in aged marrow and therefore had an unaccounted for numerical advantage, irrespective of any qualitative difference that may exist between young and old stem cells.

Additional studies have contributed to a growing consensus pointing to stem cell aging. As further examples, Rebel et al. [59] compared murine fetal liver and young adult marrow stem cells in a limiting dilution repopulation assay and found an extensive functional advantage of the fetal cells. Similarly, Chen et al. used modeling techniques based on Poisson probabilities to analyze repopulation by paired admixtures of fetal liver, young bone marrow, or old bone marrow [60]. They found an inverse correlation between donor age and functional repopulating capacity in BALB/cBy mice. Fetal liver stem cells had up to three times the functional capacity of young marrow stem cells, which, in turn, had twice the capacity of stem cells from marrow of 2-year-old animals. Morrison et al. showed that despite as much as a five-fold increase in stem cells in old marrow, the sorted cells had only about one-fourth the engrafting potential when transplanted [46].

Human and mouse stem cells sorted to high purity from hematopoietic tissue of different chronologic age including fetal liver, umbilical cord blood, and bone marrow were tested for their ability to proliferate and differentiate *in vitro*. Candidate stem cells were cultured under growth conditions in which the generation of progenitors and mature cells could be quantitated. As in repopulation studies *in vivo*, functional capacity was inversely related to chronologic age of the donor [61].

3.3.   Genetic regulation

To add to the complexity, it is now known that significant genetic differences exist between mouse strains that affect age-related changes in stem cell populations, making it difficult to compare results obtained by groups using different strains. For instance, we have compared C57BL/6 (B6) and DBA/2 (DBA) stem cell numbers and their cell cycle kinetics from fetal stages to 20-month-old adults [62]. Two significant changes were apparent. First, as has generally been found in other tissues during aging, the fraction of S-phase progenitor cells steadily decreased until at about 1 year it was below the sensitivity of the assay for both strains. At all earlier time points when it was measurable, as much as a ten-fold higher proportion of DBA cells was in S-phase. Second, stem cell numbers in both strains increased from fetal stages to young adulthood and to middle age (1 year). However, between 12 and 20 months a strain-specific difference was revealed. Whereas the number of B6 stem cells continued to increase at the same rate, the number of DBA stem cells declined by about half, to levels found in fetal liver. The relationship between these findings and the established longevity differences between these strains is taken up in a following section.

Stem cells rely heavily on interactions with stroma *in vivo* for a number of extrinsic signals favoring their survival and affecting both their proliferation and differentiation. Effects on stroma are therefore important in any considerations of age-related changes in stem cells. Although not as extensively studied as stem cells, there is evidence that cells of the stroma undergo a decline in proliferative capacity in aging that is functionally matched by a decline in their capacity to support of stem cells [63].

4. Telomeres

4.1. Relationship to replicative senescence

An important experimental tool that has lent support to an intrinsic mechanism being related to replicative exhaustion of stem cells is the analysis of telomeres. The reader is referred to extensive reviews of this expanding field [64, 65]. In brief, telomeres are the nucleotide repeats ($TTAGGG_n$ in mammals) that cap and stabilize the ends of chromosomes, preventing chromosomal fusions. They also solve the end--duplication problem of DNA synthesis identified by Watson and Olovnikov by providing the extension scaffold on which the DNA synthetic machinery can initiate complete replication of the genome [66, 67]. In so doing, telomeres themselves shorten with each cell division. Thus, telomeres provide, from their length, a replicative history of the cell and thus serve as a mitotic clock. The studies by Hayflick in the early 1960s showed that human fibroblasts have limited lifespans of roughly 50 doublings in culture [68], and it is now known that when telomeres shorten to a minimum critical length, cells become senescent, as did mitotically exhausted fibroblasts in Hayflick's cultures [69]. A few cell types are able to synthesize telomerase, an enzyme which lengthens telomeres, but most normal somatic cells, including fibroblasts, are unable to do so [70]. It has been convincingly shown that overexpression of telomerase in normal cultured cells allows them to escape replicative senescence apparently permanently, but without becoming tumorigenic [71]. These findings firmly link preservation of telomeres with immortality, at least *in vitro*.

4.2. Stem cell telomeres

There is now ample evidence that hematopoietic stem cells produce telomerase, but apparently not enough to prevent telomere erosion after extensive proliferation [46].

Vaziri et al. showed that the length of telomeres of purified stem cells varied inversely with the age of the donor [72]. Human stem cells from older adults had significantly shorter telomeres than their counterparts purified from umbilical cord blood, which, in turn, had telomeres shorter than from fetal liver stem cells. This and other data showing shortening of telomeres in blood leukocytes throughout adult life of humans [73], may suggest that telomeres play a role in the age-related decline in stem cell function discussed above, and possibly limit longevity through compromise of critical organ systems such as hematopoiesis. In further support of this notion, Rudolph et al. found that erosion of telomeres in hepatic cells of telomerase knockout mice contributed to end-stage organ failure in experimental cirrhosis [74]. Importantly, when adenovirus-mediated gene therapy restored telomerase activity in liver cells, cirrhosis was alleviated and liver function was improved.

However, other data argue that telomere length does not play an important role in organismal longevity. Perhaps most persuasive is the fact that laboratory mice have dramatically longer telomeres than do other mammalian species, including humans, and yet are relatively short-lived [75]. Moreover, it is difficult to see how the very slow rate of telomere loss per cell division (50-200 bp), and the relatively modest variation telomere repeat length amongst over 500 human subjects ranging from childhood to very old [73], could significantly impact stem cell aging.

4.2.1. *Effect of stem cell transplantation on telomere length*

In the transplantation setting, several studies of telomere length have been made in blood cells of allografted patients and their stem cell donors. Telomeres were shorter in blood cells produced by transplanted stem cells than in blood cells originating from the same stem cell pool, but remaining in the donor [76, 77]. These data suggest that the telomere differences reflect the replicative stress undergone by the transplanted stem cells during engraftment. Moreover, the size of the telomere discrepancy was inversely proportional to the number of stem cells transplanted [78]. Thus the overall replicative stress on the population was influenced by the number of stem cells participating in engraftment. The fewer the stem cells in the graft, the greater the replicative stress on the stem cell population, and vice versa. Subsequent studies have shown that the major portion of telomere shortening occurs during the first year after transplant, coinciding with the replicative stress associated with engraftment [79]. Thereafter, telomere shortening occurs at the same rate as in normally aging populations. The data on transplantation might imply that telomere shortening was instrumental in diminishing the function of the stem cell population. However, the degree of shortening accompanying engraftment is usually only about one kb in telomere length, an amount difficult to reconcile with an important cause of stem cell senescence [76].

Nonetheless, it should be pointed out that virtually all of the allograft donors studied to date have been young, usually in their teens or 20s (see for example Mathioudakis et al. [76]). The use of stem cell transplantation is continuously being expanded to treat older patients, and older marrow donors are being used for allografting. It is conceivable that the replication associated with engraftment may be considerably more stressful on an old stem cell pool than on a young one. Moreover, pairing older marrow donors with young recipients who have a lifetime of hematopoiesis ahead of them, could conceivably be problematic.

4.3.   Telomeres and aging

Although telomere length is now known to limit replicative potential of cells *in vitro*, the notion of its regulatory role in organismal longevity has been seductive but unproven in complex organisms such as mammals. Nonetheless, the disparity between lifespan and telomere length in mammals argues persuasively that the two are unrelated [75]. Much of our recent knowledge concerning the physiological role of telomeres *in vivo* has come from mice null for telomerase. Possibly due to inherently long telomeres of laboratory mice, overt pathology was not detectable in the first several generations of telomerase null animals and, importantly, they lived normal lifespans [80]. However, the sixth generation was unable to reproduce due to a lack of functional germ cells. Closer examination of this and earlier generations revealed subtle but significant changes [81]. These were especially noteworthy in the continuously renewing tissues in which stem cells were the developmental cellular source of mature cells. In cells of several tissues, karyotypic analyses demonstrated potential sources of genetic instability in the form of extensive chromosomal abnormalities, including fusions.

Detailed longitudinal studies of telomerase knockout mice have now shown that particularly sixth-generation animals have shortened lifespans and prematurely undergo physiological changes associated with aging in some but not all tissues [81]. As in aged individuals, homeostatic functioning of most organ systems remains intact, but responses to physiologic stresses such as wound healing and hematopoietic recovery after chemotherapy is dramatically blunted. Thus, it appears that regenerative reserves in renewing tissues may be limited through effects on their respective stem cell populations.

5.   A genetic link between cell replication, stem cell population size, and organismal lifespan in the mouse

5.1.   Study of chimeric mice

Our lab has been studying interstrain differences in the hematopoietic populations of mice, particularly as they relate to aging. These studies were prompted by the observation that early hematopoietic progenitor cells of different mouse strains varied widely with respect to cell cycle kinetics. The spleen colony-forming cell (CFU-S) populations of young adult mice of the B6 and DBA strains normally differed by ten-fold with respect to the population fraction in S-phase under steady-state conditions [82]. Yet these strains have very similar mature blood cell counts, suggesting that regulatory checkpoints exist at multiple levels in the developmental hierarchy. Chimeras (allophenic mice) were constructed by combining preimplantation-stage embryos of the two strains in order to determine if these and possibly other cryptic or subtle hematologic differences were cell-autonomous. Extrinsically regulated differences in stem cell function would be obviated in chimeras where stem cells of the two strains share a common environment.

Our principal interest was in uncovering genetic differences manifested in the developmentally early cells themselves since they would certainly provide better insight into stem cell biology and may shed light on stem cell diseases, including leukemia. Initial studies of these chimeras revealed strain-specific susceptibility to leukemia induced by Friend murine leukemia virus [83], and a still unexplained skewing in uninfected animals of blood cell lineages [82]. When normalized for overall chimerism in the hematopoietic system, DBA red cell and platelet formation was disproportionately high and neutrophil and especially lymphocyte production was skewed low. Long-term studies of chimerism in the blood formation were carried out because it was anticipated that subtle, cell-autonomous differences in the organization of the respective stem and progenitor populations might appear only after many months or in aged animals.

Another impetus for long-term studies of chimeras was the knowledge that DBA was a relatively short-lived strain and B6 was a long-lived strain. Invoking Hayflick's cell mitotic limit, we wondered if the rapidly proliferating DBA stem cell population might be replicatively exhausted *in vivo* and in some way contribute to the shorter lifespan of this strain. If the cell cycle phenotype was cell-autonomous it might be manifested in chimeras by a cessation of hematopoiesis derived from the DBA stem cells and blood formation would become wholly B6 in origin. This is in fact what was found in most chimeras [31]. After about a year of stable blood cell chimerism, DBA representation in blood cells began to diminish and in many animals was completely absent after 2 years — approximately the normal the lifespan of this strain. Chimeras did not die at this time because the longer-lived B6 strain took over all blood cell production.

5.2.   Experimental reversal of stem cell activation and quiescence

We next asked whether the DBA stem cells had disappeared completely, or whether some or all had entered a state of quiescence. To answer this, we harvested marrow from chimeras whose blood cells had become entirely B6 in origin, and transplanted it into irradiated recipients to determine the origin of engraftment. Surprisingly, cryptic DBA stem cells initially contributed heavily to engraftment but, after several months became completely quiescent again [84]. Subsequent transplantation of marrow from primary to secondary recipients revealed the same pattern: DBA stem cells were reactivated and participated in blood formation, albeit to a lesser extent than in primary hosts. Again, after an even shorter period they became quiescent and all blood cells were of B6 origin. The fact that the limited extent of DBA chimerism and its short duration after secondary transplant argued for a DBA-specific decline in stem cell function. It was now clear that at least some stem cell phenotypes were cell -- autonomous or intrinsic, thus opening the way for genetic studies.

5.3.   Genetic studies

5.3.1.   *Stem cell assay in vitro*

A reliable and sensitive assay for stem cells was needed in order to embark upon genetic studies to map, clone and characterize genes responsible for the observed stem cell phenotypes. Since screening large numbers of mice would be required for the next step, an *in vitro* assay was chosen for practical reasons and cost considerations. The assay was based on the long-term bone marrow cell culture technique originally developed by Dexter and Lajtha [85], and subsequently refined into a quantitative assay by Cashman et al. [86], and Ploemacher et al. [87]. The cobblestone area forming cell (CAFC) assay takes advantage of the fact that stem cells require a close association with stromal cells in order to survive and function. To this end, limiting dilutions of a stem cell source are added to cultures of established stromal cell monolayers. We used a mouse line, FBMD-1, but primary stroma established from hematopoietic organs, or other cell lines work as well. Individual stem cells translocate beneath the stroma and after a period of quiescence, proliferate to form small colonies (5-50 cells) that appear as cobblestones under phase contrast illumination. The length of the period of quiescence is variable for CAFCs and it has been shown by Ploemacher's group that quiescence is directly proportional to the developmental "primitiveness" of the stem cell [87]. In this way subpopulations of stem cells within a hierarchy can be delineated: the less primitive are more numerous, have a higher fraction in S-phase and form colonies in about a week. The most primitive have ~100-fold lower frequency, are largely out of cell cycle, and are quiescent for more than a month before forming colonies. Single cell sorting experiments have shown that CAFCs form colonies only once and at a time commensurate with their position in the hierarchy.

To obtain comparative inter-strain data, we first measured the numbers of CAFCs on days 7, 14, 21, 28, and 35 in the bone marrow of eight commonly used inbred mouse strains. Whereas the numbers of day 7 cells was nearly uniform, numbers of day 35 cells varied by nearly ten-fold amongst mouse strains [58, 88]. As we had seen before using the CFU-S assay, the fraction of day 7 CAFCs in S-phase also varied by more than ten-fold amongst strains. Only AKR mice had a detectable number of the more primitive CAFC-day 35 in S-phase and, interestingly, these mice are very short-lived and die of lymphoma.

5.3.2. *Average lifespan correlates with cell cycle kinetics*

Surprisingly, when CAFC-day 7 cycling was plotted against the mean lifespans of the eight strains, a highly significant inverse correlation was apparent such that nearly two--thirds of the variability in lifespan amongst strains was accounted for by differences in the cycling phenotype [58, 88]. These findings reinforced our hypothesis derived from the study of chimeras, that the two phenotypes were closely related, had a strong genetic component, and were amenable to genetic analysis.

B6 and DBA mice were chosen for further genetic analysis for several reasons. First, they had been the subject of our previous chimeric work. Second, using the CAFC assay the two strains consistently displayed nearly the largest differences in cycling, stem cell numbers, and lifespan. Third, powerful genetic resources were available to study these strains. Most significantly, a set of 26 (recently expanded to 34) recombinant inbred (BXD) mouse strains had been established at The Jackson Laboratory and were commercially available. As a result of genetic studies in these strains over the past 20 years by a large number of labs studying many genetic traits, over 2000 loci and molecular markers have been mapped in these strains. Moreover, advanced intercrosses are ongoing between the two strains that now exceed ten generations (Robert Williams, pers. comm.). Advanced intercross mice, because of the greatly increased genetic recombination, provide a tool for higher resolution mapping.

5.3.3.   *Linkage analysis in recombinant inbred mouse strains*

Consequently, we phenotyped the BXD strains for cell cycling and CAFC numbers [88]. We were particularly interested in the relationship we had found previously between cycling and lifespan in the eight unrelated or distantly related inbred strains. To this end, we found that the difference in cycling between B6 and DBA was essentially accounted for by four quantitative trait loci (QTL), major loci on chromosomes 7 and 11, and minor loci on chromosomes 4 and 9. Fortuitously, Gelman et al. [89] had carried out longevity studies on the BXD strains in the late 1980s. By plotting their longevity data against our cell cycling data, we found the same inverse correlation that we had found with genetically diverse inbred mice; that is, the higher the cycling rate, the shorter the lifespan. At the time the longevity studies were published, the genetic map was not sufficiently detailed nor were sophisticated linkage analysis tools available for accurate mapping of QTL.

5.3.4.   *Quantitative trait loci affecting lifespan and cell cycle kinetics map to the same genomic locations*

We therefore re-analyzed the BXD lifespan data and found that six QTL accounted for essentially all of the natural variation in lifespan between B6 and DBA: major loci on chromosomes 2, 4, 7 (three loci), and 11 [62]; Geiger et al., unpublished). The loci on chromosomes 7 and 11 showed linkage to the same genomic markers as did the QTL regulating CAFC cycling. Thus, two of the four QTL contributing to each trait mapped to the same genomic intervals, suggesting, but not proving, a cause-and-effect relationship.

Similarly, linkage analysis of QTL affecting the sizes of the hematopoietic stem and progenitor cell populations in young and old BXD mice mapped to the same genomic intervals on chromosomes 2 and 7 as QTL determining longevity (Geiger et al., unpublished). Thus in sum, five of the six lifespan QTL had similar map locations as QTL regulating aspects of the hematopoietic stem/progenitor cell system.

5.3.5.   *Generation of congenic mice to create mouse models for future studies*

As a first step toward identifying and studying the function of genes affecting longevity in mice, we have now moved, by backcrossing, each of the genomic locations of interest from B6 to DBA/2, and vice versa, to generate congenic mouse strains. These mice are again genomic chimeras, but only in a very restrictive sense: they have donor strain genes only in the prescribed genomic segment of interest. All other genomic loci are homozygous for recipient strain alleles. We have used a 'speed' congenic strategy in which breeders for each generation are selected not only for the presence of the appropriate derivation of the interval of interest, but also for the least heterozygosity throughout the rest of the genome [90, 91]. This genomic marker-assisted approach reduces to four or five the number of backcross generations required to reach a genetic background less than 1% heterozygous. Thus, with these congenic mice it is possible to determine the effect of individual, selected genes on a trait in the context of a fixed genetic background. For example, it is anticipated that two effects will be seen in the B6 congenic line bearing the DBA/2 loci that affect both cell-cycling and longevity: the cell cycling of hematopoietic progenitors will increase to the level of DBA/2 mice, and the lifespan of the mice will decrease. Phenotypic study of these congenic mice has borne out the first prediction that a higher cell cycling has been conferred; longevity data will not be available for some time.

Thus, an important determinant in mammalian longevity appears to be the rate of cellular proliferation in key tissues *in vivo*. We have identified progenitor cells of the hematopoietic system as one of these and at present we don't know if this cell population is unique in this regard, or if they are reflective of a genetically determined proliferation rate in other tissues as well. If so, other continuously renewing tissues such as the liver, skin and the intestinal lining are attractive candidates, although genetically determined cell-cycling rates may be reflected in an even more widespread spectrum of tissues. Given the identification of telomeres as the long-sought 'mitotic clock', that sets the Hayflick limit for cells in culture [92], it is appealing to invoke telomeres mechanistically in the context of longevity of the organism. It is certainly true that proliferative tissues, in contrast with post-mitotic ones, are more prone to genetic damage [93]. However, further studies are needed to further develop, or disprove, a link between telomeres, cell replication, and longevity.

It remains to be seen whether or not the common genes that regulate cell proliferation and lifespan of mice have homologues with similar functions in organisms whose adult tissues are all essentially post-mitotic, such as the nematode *Caenorhabditis elegans*. Conversely, it remains to be seen whether genes that contribute to lifespan extension in such non-mammalian animals affect longevity in mammals.

6.   Conclusions and future directions

Genetic determinants of longevity have proven surprisingly straightforward to unravel in yeast, flies, worms, and even in mammals [94, 95]. Using the limited context of lifespan differences between selected mouse strains, a handful of QTL account for at least the major differences. Given the advances in genetic analysis in the last few years and with whole genome sequences soon to be available for human and mouse, the prospects for cloning longevity affecting genes are promising. These heady advances in genetic analysis notwithstanding, it should be remembered that the only present manipulation that can increase longevity, albeit modestly, in any mammal is restricting caloric intake [96].

The finding that cell cycle kinetics of hematopoietic progenitor cells are related to organismal lifespan in mice is provocative since it provides at least a conceptual link between replicative senescence *in vitro* and organismal lifespan [97]. It remains to be seen whether or not hematopoietic progenitor cell cycling is the sole population affected by the putative loci and therefore that the fate of these cells affects longevity, or if it is merely the only cell population analyzed to date. It is possible that the QTL affecting cycling in this population have similar effects on other stem cell populations that may (too) be lifespan limiting. Since maintenance of telomere length in cultured cells delays or prevents replicative senescence, it is reasonable to suggest that telomere shortening may be involved in the decline of certain tissues *in vivo* as well. In support of this notion, studies of telomerase knockout mice have shown that it is the renewing adult tissues -- presumably their respective stem cells -- that are most affected. Whether or not QTL affecting cell cycling and lifespan specify some aspect of telomere regulation remains to be seen.

7.   References

1.   Potten, C. S. Stem Cells. (1997) London, Academic Press.
2.   Kay, H. E. M. (1965) Lancet ii, 418
3.   Osgood, E. E. (1957) J. Natl. Cancer Inst. 18, 155-166
4.   Beddington, R. S. P. and Robertson, E. J. (1999) Cell 96, 195-209
5.   Till, J. E.; McCulloch, E. A. and Siminovitch, L. (1964) Proc. Natl. Acad. Sci. USA 51, 29-36
6.   Weglarz, T. C. and Sandgren, E. P. (2000) Proc. Natl. Acad. Sci. USA 97, 12595-12600
7.   Visser, J. W.; Bauman, J. G.; Mulder, A. H.; Eliason, J.F. and de Leeuw, A. M. (1984) J Exp Med 159, 1576-90
8.   Spangrude, G. J.; Heimfeld, S. and Weissman, I. L. (1988) Science 241, 58-62
9.   Ploemacher, R. E. and Brons, N. H. C. (1988) Exp. Hematol. 16, 21-26
10. Uchida, N. and Weissman, I. L. (1992) J. Exp. Med. 175, 175-184
11. Dick, J. E. , Lapidot, T., Vormoor, J., Larochelle, A., Bonnet, D., and Wang, J. (1995) Ontogeny of Hematopoiesis (Gluckman, E. and Coulombel, L., eds) pp. 97-101, Editions Inserm,
12. Morrison, S. J.; White, P. M.; Zock, C. and Anderson, D. J. (1999) Cell 96, 737-749
13. Uchida, N.; Buck, D. W.; He, D.; Reitsma, M. J.; Masek, M.; Phan, T. V.; Tsukamoto, A. S.; Gage, F. H. and Weissman, I.L. (2000) Proc. Natl. Acad. Sci. USA 97, 14720-14725
14. Spangrude, G. J.; Brooks, D. M. and Tumas, D. B. (1995) Blood 85, 1006-1016
15. Osawa, M.; Hanada, K.; Hamada, H. and Nakauchi, H. (1996) Science 273, 242-245
16. Siminovitch, L.; Till, J. E. and McCulloch, E. A. (1964) J. Cell. Comp. Physiol. 64, 23-31
17. Harrison, D. E. (1972) Nature New Biology 237, 220-222
18. Szilvassy, S. J.; Humphries, R. K.; Lansdorp, P. M.; Eaves, A. C. and Eaves, C. J. (1990) Proc. Natl. Acad. Sci. USA. 87, 8736-8740
19. Fraser, C. C.; Eaves, C. J.; Szilvassy, S. J. and Humphries, K. R. (1990) Blood 76, 1071-1076
20. Smith, L. G.; Weissman, I. L. and Heimfeld, S. (1991) Proc. Natl. Acad. Sci. USA. 88, 2788-2792
21. Iscove, N. N. and Nawa, K. (1997) Curr. Biol. 7, 805-808
22. Jordan, C. T. and Lemischka, I. R. (1990) Genes Dev. 4, 220-232
23. Capel, B.; Hawley, R.; Covarrubias, L.; Hawley, T. and Mintz, B. (1989) Proc. Natl. Acad. Sci. USA. 86, 4564-4568
24. Capel, B.; Hawley, R. G. and Mintz, B. (1990) Blood 75, 2267-2270
25. Nikolic, B. and Sykes, M. (1997) Immunol. Res. 16, 217-228

26. Keller, G. and Snodgrass, R. (1990) J. Exp. Med. 171, 1407-1418
27. Harrison, D. E.; Astle, C. M. and Delaittre, J. A. (1978) J. Exp. Med. 147, 1526-1531
28. Ross, E.; Anderson, N. and Micklem, H.S. (1982) J. Exp. Med. 155, 432-444
29. Harrison, D. E.; Lerner, C.; Hoppe, P. C.; Carlson, G. A. and Alling, D. (1987) Blood 69, 773-777
30. Warner, C. M.; McIvor, J. L. and Stephens, T. J. (1977) Differentiation 9, 11-17
31. Van Zant, G.; Holland, B. P.; Eldridge, P. W. and Chen, J.-J. (1990) J. Exp. Med. 171, 1547-1565
32. Bradford, G. B.; Williams, B.; Rossi, R. and Bertoncello, I. (1997) Exp. Hematol. 25, 445-453
33. Cheshier, S. H.; Morrison, S. J.; Liao, X. and Weissman, I. L. (1999) Proc. Natl. Acad. Sci. USA 96, 3120-3125
34. Pietrzyk, M. E.; Priestley, G. V. and Wolf, N. S. (1985) Blood 66, 1460-1462
35. Abkowitz, J. L.; Golinelli, D.; Harrison, D. E. and Guttorp, P. (2000 Blood 96, 3399-3405
36. Abkowitz, J. L.; Lineberger, M. L.; Newton, M. A.; Shelton, G. H.; Ott, R. L. and Guttorp, P. (1990) Proc. Natl. Acad. Sci. USA. 87, 9062-9066
37. Gale, R.E.; Fielding, A.K.; Harrison, C.N. and Linch, D.C. (1997) Br.J.Haematol. 98, 512-519
38. Micklem, H.S.; Lennon, J.E.; Ansell, J.D. and Gray, R.A. (1987) Exp. Hematol. 15, 251-257
39. Buller, R. E.; Sood, A. K.; Lallas, T.; Buekers, T. and Skilling, J. S. (1999) J. Natl. Cancer Inst. 91, 339-346
40. Abkowitz, J. L.; Taboada, M.; Shelton, G. H.; Catlin, S. N.; Guttorp, P. and Kiklevich, J. V. (1998) Proc. Natl. Acad. Sci. USA 95, 3862-3866
41. Vaux, D. L. and Korsmeyer, S. J. (1999) Cell 96, 245-254
42. Domen, J.; Gandy, K. L. and Weissman, I. L. (1998) Blood. 91, 2272-2282
43. Domen, J.; Cheshier, S. H. and Weissman, I. L. (2000) Journal of Experimental Medicine 191, 253-263
44. Potten, C. S. and Loeffler, M. (1990) Development. 110, 1001-1020
45. Winton, D. J. and Ponder, B. A. (1990) Proc. Royal Soc. Lond. B Biol. Sci. 241, 13-18
46. Morrison, S. J.; Wandycz, A. M.; Akashi, K.; Globerson, A. and Weissman, I. L. (1996) Nature Med. 2, 1011-1016
47. Globerson, A. (1999) Exp. Gerontol. 34, 137-146
48. Williams, L. H.; Udupa, K. B. and Lipschitz, D. A. (1986) Expp. Hematol. 14, 827-832
49. Clark, M. R. (1988) Physiol. Rev. 68, 503-554
50. Miller, R. A. (1996) Science 273, 70-74
51. Effros, R. B. (1998) Am. J. Hum. Genet. 62, 1003-1007
52. Ogden, D. A. and Micklem, H. S. (1976) Transplantation 22, 287-293
53. Micklem, H. S.; Ford, C. E.; Evans, E. P.; Ogden, D. A. and Papworth, D. S. (1972) J. Cell Physiol. 79, 293-298
54. Mauch, P.; Botnick, L. E.; Hannon, E. C.; Obbagy, J. and Hellman, S. (1982) Blood 60, 245-252
55. Harrison, D. E. (1983) J. Exp. Med. 157, 1496-1504
56. Harrison, D. E. (1980) Blood 55, 77-81
57. Harrison, D. E.; Astle, C. M. and Stone, M. (1989) J. Immunol. 142, 3833-3840
58. de Haan, G.; Nijhof, W. and Van Zant, G. (1997) Blood. 89, 1543-1550
59. Rebel, V. I.; Miller, C. L.; Eaves, C. J. and Lansdorp, P. M. (1996) Blood. 87, 3500-3507
60. Chen, J.; Astle, B. A. and Harrison, D. E. (1999) Exp. Hematol. 27, 928-935
61. Lansdorp, P.M.; Dragowska, W. and Manyani, H. (1993) J. Exp. Med. 178, 787-791
62. de Haan, G. and Van Zant, G. (1999) FASEB J. 13, 707-713
63. Boggs, S. S.; Patrene, K. D.; Austin, C. A.; Vecchini, F. and Tollerud, D. J. (1991) Exp. Hematol. 19, 683-687

64. Blackburn, E. M. and Greider, C. W. Telomeres. (1995) Cold Spring Harbor, NY, Cold Spring Harbor Laboratory Press.

65. Chadwick, D. J. and Cardew, G. Telomeres and Telomerase. (1997) New York, John Wiley and Sons.

66. Olovnikov, A. M. (1971) Doklady Akad. Nauk. SSSR 201, 1496-1499

67. Watson, J. D. (1972) Nature (New Biology) 239, 197-201

68. Hayflick, L. and Moorhead, P. S. (1961) J. Exp. Cell Res. 25, 585-621

69. Harley, C. B.; Futcher, A. B. and Greider, C. W. (1990) Nature 345, 458-460

70. Liu, K.; Schoonmaker, M. M.; Levine, B. L.; June, C. H.; Hodes, R. J. and Weng, N. (1999) Proc. Natl. Acad. Sci. USA 96, 5147-5152

71. Bodnar, A. G.; Ouellette, M.; Frolkis, M.; Holt, S. E.; Chiu, C.-P.; Morin, G. B.; Harley, C. B.; Shay, J. W.; Lichtsteiner, S. and Wright, WE. (1998) Science 279, 349-352

72. Vaziri, H.; Dragowska, W.; Allsopp, R. C.; Thomas, T. E.; Harley, C. B. and Lansdorp, P. M. (1994) Proc. Natl. Acad. Sci. USA. 91, 9857-9860

73. Rufer, N.; Brummendorf, T. H.; Kolvraa, S.; Bischoff, C.; Christensen, K.; Wadsworth, L.; Schulzer, M. and Lansdorp, P. M. (1999) Journal of Experimental Medicine 190, 157-167

74. Rudolph, K. L.; Chang, S.; Millard, M.; Schreiber-Agus, N. and DePinho, R. A. (2000) Science 287, 1253-1258

75. Hemann, M.T. and Greider, C.W. (2000) Nucleic Acids Res. 28, 4474-4478

76. Mathioudakis, G.; Storb, R.; McSweeney, P. A.; Torok-Storb, B.; Lansdorp, P. M.; Brummendorf, T. H.; Gass, M. J.; Bryant, E. M.; Storek, J.; Flowers, M. E. D.; Gooley, T. and Nash, R. A. (2000) Blood 96, 3991-3994

77. Wynn, R. F.; Cross, M. A.; Hatton, C.; Will, A. M.; Lashford, L. S.; Dexter, T. M. and Testa, N. G. (1998) Lancet 351, 178-181

78. Notaro, R.; Cimmino, A.; Tabarini, D.; Rotoli, B. and Luzzatto, L. (1997) Proc. Natl. Acad. Sci. USA. 94, 13782-13785

79. Rufer, N.; Brummendorf, T. H.; Chapuis, B.; Heig, C.; Lansdorp, P. M. and Roosnek, E. (2001) Blood 97, 575-577

80. Blasco, M. A.; Lee, H. W.; Hande, M. P.; Samper, E.; Lansdorp, P. M.; DePinho, R. A. and Greider, C. W. (1997) Cell 91, 25-34

81. Rudolph, K. L.; Chang, S.; Lee, H.-W.; Blasco, M.; Gottlieb, G. J.; Greider, C. W. and DePinho, R. A. (1999) Cell 96, 701-712

82. Van Zant, G.; Eldridge, P. W.; Behringer, R. R. and Dewey, M. J. (1983) Cell 35, 639-645

83. Eldridge, P. W., and Dewey, M. J. (1986) Exp. Hematol. 14, 380-385

84. Van Zant, G.; Scott-Micus, K.; Thompson, B. P.; Fleischman, R. A. and Perkins, S. (1992) Exp. Hematol. 20, 470-475

85. Dexter, T. M. and Lajtha, L. G. (1974) Br. J. Haematol. 28, 525-530

86. Cashman, J.; Eaves, A. C. and Eaves, C. J. (1985) Blood 66, 1002-1005

87. Ploemacher, R. E.; van der Sluijs, J. P.; van Beurden, C. A.; Baert, M. R. and Chan, P. L. (1991) Blood 78, 2527-2533

88. de Haan, G. and VanZant, G. (1997) J. Exp. Med. 186, 529-536

89. Gelman, R.; Watson, A.; Bronson, R. and Yunis, E. (1988) Genetics 118, 693-704

90. Markel, P.; Shu, P.; Ebeling, C.; Carlson, G. A.; Nagle, D. L.; Smutko, JS. and Moore, KJ. (1997) Nat. Genet. 17, 280-284

91. Wakeland, E.; Morel, L.; Achey, K.; Yui, M. and Longmate, J. (1997) Immunol.Today 18, 472-477

92. de Lange, T. (1998) Science 279, 334-335

93. Dolle, M.; Giese, H.; Hopkins, C. L.; Martus, H.-J.; Hausdorff, J. M. and Vijg, J. (1997) Nat. Genet. 17, 431-434

94. Guarente, L. and Kenyon, C. (2000) Nature 408, 255-262

95. Rogina, B.; Reenan, R. A.; Nilsen, S. P. and Helfand, S. L. (2000) Science 209, 2137-2140

96. Sprott, R. L. (1997) Exp. Gerontol. 32, 205-214

97. Campisi, J. (1996) Cell 84, 497-500

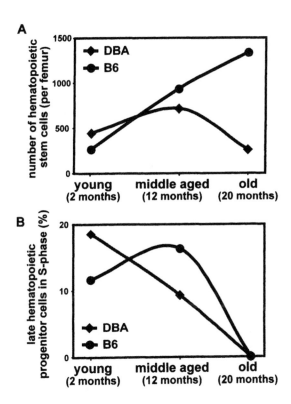

Figure 1. Changes in number of hematopoietic stem cells and the cycling status of hematopoietic progenitor cells in DBA/2 (DBA) and C57BL/6 (B6) mice during aging. A) The number of hematopoietic stem cells was determined with the CAFC assay (CAFC day 35) at different timepoints. Differences between B6 and DBA values are significant for all timepoints measured. B) The percentage of late hematopoietic progenitor cells in S-phase (CAFC day 7) was measured at different timepoints. Differences between B6 and DBA values are significant for all timepoints measured, except for 20 months.

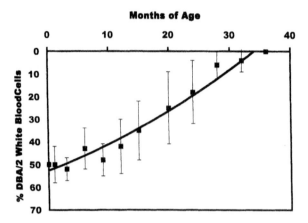

Figure 2.   Percent DBA/2 leukocytes in peripheral blood of C57BL/6←→DBA/2 chimeric mice. The chimerism of white blood cells in embryo-aggregated chimeric mice was monitored over three years. The means plus/minus the standard deviations of 6-8 mice are presented at each time point. Since each chimera has a unique initial level of chimerism (due to random contributions of blastomeres of the C57BL/6 and DBA/2 co-joined blastulae to the inner cell mass destined to become the embryo proper), the initial chimerism for each was normalized to a common 50% value initially. Note that contributions of DBA/2 stem cells decline steadily with time. During the third year of life most chimeras have only C57BL/6 blood cells.

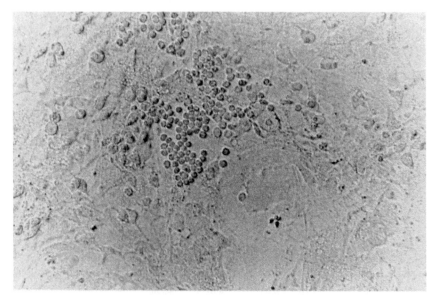

Figrure 3.   Photograph of cobblestone area growing beneath a monolayer of stromal cells. The cobblestone shown here was derived from a very primitive hematopoietic cell that only proliferated to form the visible cobblestone after five weeks in culture.

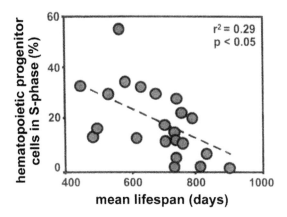

Figure 4. Progenitor cell cycling and mean lifespan are inversely correlated in BXD recombinant inbred mice. A significant negative correlation was observed between the mean lifespan and the percentage of late hematopoietic progenitor cells in S-phase of 22 BXD strains.

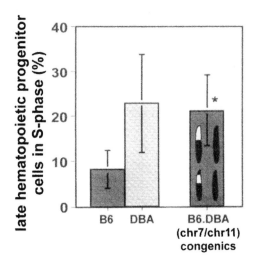

Figure 5. Congenic animals confirm the linkage analysis. Congenic animals were generated with a speed congenic approach. Both intervals that contain the loci linked to cycling (chromosome 7 and chromosome 11) were moved from a DBA/2 background onto a C57BL/6 background. Shown is the analysis for the percentage of late HPCs in S-Phase. Note that the phenotype moves with the genotype, confirming our linkage analysis. $p < 0.05$ for B6 *versus* B6.DBA (chr7/chr11).

Table 1.   Comparative linked locus list for multiple quantitative trait mapping in the BXD recombinant inbred set.

Linked locus		Trait			
				young animals (<2 months)	old animals (>18 months)
		lifespan	cycling of IHPCs	eHPCs/femur	HSCs/femur
chr 1	(38 CM)			B++	
chr 2	(70 CM)	D+			B+++
chr 4	(42 CM)	D+			
chr 7	(1 CM)	B+++	D++		
	(8 CM)	B+		D+	
	(25 CM)	B+		D+	
	(50 CM)			D++	
chr 11	(28 CM)	B+	D+++		
chr 18	(27 CM)				D++

Linkage analysis was performed for the traits lifespan, percentage of late hematopoietic progenitor cells (IHPCs) in cycle, number per femur of early hematopoietic progenitor cells (eHPCs) and hematopoietic stem cells (HSCs) in young (<3 months) and old (>18 months) animals. In the first column all linked loci that showed linkage with an at least suggestive linkage value are listed with the centimorgan (cM) position of the marker with the highest significance in the interval. The type of capital letter indicates which segment (either DBA/2 (D) or C57BL/6 (B) derived) is responsible for a higher trait value (lifespan: longer lifespan). For all linked loci the suggestive, significant and highly significant LRS values were individually calculated: +: suggestive, ++: higher than suggestive, and up to significant and +++: above significant

CHAPTER 3

HAEMATOPOIETIC STEM CELLS AND THE THYMUS IN AGING

AMIELA GLOBERSON

Table of contents

Stem Cells: A Cellular Fountain of Youth.   Ed. by Mark P. Mattson and Gary Van Zant. 43 — 72

1.   Introduction

The T lymphocytes develop from extra thymic stem cells that migrate from hematopoi-etic tissues and settle in the thymus. Considerable progress has been accomplished in research on hematopoiesis and the ontogeny of thymocytopoiesis, as reviewed at length [1-4], and aspects related to aging have also been documented [5, 6]. It has become clear that T lymphocyte development in the thymus depends on a functional microenvironemt. Furthermore, it was noted that cross interactions between the various thymocyte subsets and stromal components may determine the outcome of the structural changes in the microenvironment and eventual T cell development [7, 8]. Whereas most of the studies focused on the nature of stem cells that differentiate into thymic lymphocytes and on their fate in aging. there have been also some efforts to elucidate the status of stem cells that give rise to the microenvironmental compartments in the thymus. An account of these effects of aging on the hematopoietic stem cells - thymus axis has been reported recently [5, 6, 9].

Age-related thymic involution raises several points that deserve attention. Firstly, generation of naive T lymphocytes continues through very old age, as shown in centenarians [10, 11]. Whereas these cells may have been produced in the residual aging thymus, other extra-thymic sources for T cell development in aging need to be considered. Secondly, the thymus is known for its plasticity, as exhibited in the seasonal thymic involution in reptiles [12, 14], whereas age-related thymic involution is irreversible under normal physiological conditions. The mechanisms underlying regeneration and replenishment of stem cells is therefore critical for a comprehensive understanding of the developmental potential of the thymus, and of T cell development in aging.

Hematopoietic stem cells have been catalogued by now on the basis of membrane markers (e.g. CD34), receptors to cytokines (e.g., c-kit, Sca-1), and functional potential (e.g., short and long-term reconstitution of hematopoiesis) as reviewed in detail [1-5, 15]. The various stem cell properties are beyond the scope of this chapter. The chapter screens the current knowledge on age-related changes, along the early processes of T cell development at the stem cell-thymus axis prior to the T cell receptor (TCR) dependent phase. It reviews experimental data relevant to aging, and draws attention to questions that are still open in that respect.

2.   Thymic stem cells

2.1.   The lymphoid cell compartment

The classical debate on the issue of thymic stem cells origin has been reviewed extensively [1-4, 16] and the principles are outlined here just briefly.

The fact that the thymus is the first lymphoid organ that appears in early ontogeny has led to the hypothesis that thymic stem cells originate intrinsically within this tissue. The classical studies of Auerbach [17] demonstrated by a variety of *in vivo* and *in vitro* experimental conditions that thymic rudiments from 12 day mouse embryos were independent of any extrinsic source of stem cells to develop into a lymphoid organ, in contrast to the presumptive embryonic spleen that did not give rise to lymphocytes on its own. It was thus proposed that the thymus is the primary source of lymphopoietic stem cells in ontogeny. However, subsequent studies indicated that thymic rudiments from embryos at an earlier stage (10-11 days rather than 12) failed to develop into a lymphoid organ [18]. Accordingly, it was concluded that the early stem cells in the embryonic thymus derive from extrinsic sources [19]. The anatomic site where these stem cells first emerge in the embryo, was subsequently a matter of major controversy.

Hematopoietic stem cells have been identified in the extra-embryonic yolk sac tissue [2] and the potential of the yolk sac cells to differentiate into T cells has been indicated from a variety of studies [20]. Hence, mouse embryonic yolk sac cells could elicit a graft *versus* host response [21], and natural killer (NK)-like cell function [22]. In addition, yolk sac cells differentiated *in vitro* to cells expressing T lymphocyte membrane markers [23, 24]. Whereas these studies indicate that yolk sac cells can give rise to lymphopoiesis, and that it may play a role of an effector organ in early ontogeny [20], there was no critical evidence that the yolk sac is the source of the stem cells in the embryonic thymus. On the contrary, a large volume of data on the phylogeny of lymphoid development, resulting from studies on amphibians and birds, pointed to intra-embryonic sites of relevant stem cell origins [3, 16]. Subsequent studies provided critical evidence that the intrinsic embryonic site of hematopoietic stem cells starts from the aorta-gonadal membrane (AGM) region. The AGM region was recognized as an independent site of hematopoietic cell generation in the embryo proper that appears in this region before circulation is established between the yolk sac and the embryo itself. Sequential changes were identified through the fetal liver (in mammals) to the ultimate location in the bone marrow [25]. This anatomical hierarchy is associated with differences in biological properties of the stem cells [2]. Awareness of changes in the stem cells site and properties has thus led to the question of whether the stem cells compartment continues to change throughout the lifespan of the individual. Indeed, stem cells within the bone marrow were found to change with age, as is manifested to a great extent at the T cell development level as recently reviewed [5, 6] and is further discussed in the following sections.

2.2.   The thymic microenvironment

Whereas the above studies focused on lymphohemopoietic cells, one needs to consider also the status and function of the thymic microenvironment. The classical studies of Auerbach showed very clearly that thymic lymphoid development in the embryo depends on epithelial - mesenchymal cell interaction [17], pointing to the mesenchyme as the initial source of the thymic stroma in the embryo. The progress in thymobiology and establishment of advanced methods for studies of early events in T cell development have enabled a critical approach to the role of the thymic mesenchyme [26, 27].

The classical studies of Auerbach [17] have also established the principle of cell-cell interactions underlying lymphoid development. This principle was shown also in lymphoid regeneration following irradiation of the adult thymus [28], and was proposed as a mechanism of continuous control of lymphoid development in the adult [29]. The nature and roles of the thymic microenvironment for lymphopoiesis in the thymus has been studied extensively since then. As it stands, the stroma provides non-TCR-mediated interaction with thymocyte progenitors and early immature T cells.

The embryonic thymic microenvironment derives from neural crest cells, and extirpation of this region results in impaired thymic development [27]. The source(s) and nature of stem cells that give rise to the diverse components of the microenvironment deserve further attention. The thymic microenvironment is characterized by its distinct architecture of cortex and medulla that are composed of different cell types, including epithelial cells, macrophages, dendritic and antigen-presenting cells that manifest different properties in the cortex and medulla [30, 31]. The question of whether cortical and medullary epithelial cells derive from a common, or from different stem cells, was approached by Ropte et al. [32], who identified cells that showed double-labeling for markers of both cortical and medullary epithelial cells.

Distinct age-related changes in morphology and in expression of stromal cell markers have been noted [33]. However, the basis of these changes, particularly in terms of stromal cell development, still needs to be elucidated. Whereas the above data relates to early embryonic development, it is still not clear if the thymic stroma continues to gain extra-thymic stem cells that contribute to the stromal epithelial milieu.

Other components of the thymic microenvironment include dendritic cells and macrophages that develop from extra-thymic stem cells [cf. 34].

2.3.   Stem cell commitment to T lymphocyte development

The decrease in T cell development in the thymus could be due to aging effects at the stage of commitment to lymphocytes. Studies on mechanisms that determine the commitment to T cell development have pointed to several genes that may play a role in this respect, but their function and fate in aging still need to be elucidated.

The ikaros gene family has been shown to determine commitment to both B and T lymphocytes in the bone marrow stem cell compartment [35]. Ikaros was also shown to play a role in dendritic cell development [36]. However, so far there are no data on ikaros involvement in any of these processes in relation to thymic involution.

Commitment to T cell development is dependent on Notch signaling [37-39]. Whereas T cell development was severely impaired in Notch1 deficient mice, myeloid and dendritic cells (DCs) were normal [cf. 34]. This suggests that the origin of thymic DCs is not from a common T/DC precursor, or that requirement for Notch differs for these lineages. Notch involvement was also shown for extra-thymic T cell development in the gut [40].

Notch1 is expressed on CD34+ hematopoietic precursors, and one of its ligands Jagged1 is expressed on bone marrow stromal cells. As shown by Walker et al. [41], Notch can modulate a growth factor signal (e.g., IL-3, IL6, SCF and G-CSF, compared with CSF alone), and in the absence of growth factor stimulation the Jagged1-Notch1 pathway preserves CD34+ in an immature state. Expression of Notch and/or availability of the proper cytokine signals in the aged may thus determine whether the stem cells remain immature or progress to generate T cells.

The status of these genes in aging, and whether reduced generation of T cells is causally related to these mechanisms still need to be established.

3.   Thymocytopoiesis in aging

3.1.   Naive and activated/memory T cell populations in aging

The T cell pool in the peripheral blood and lymphoid tissues of aged humans and experimental animals shows reduced levels of naive cells and an increased proportion of memory/activated cell types [42]. Whereas initially this altered proportion of cells seemed to be based on excess accumulation of memory cells on the account of newly generated cells, more recent data point to more complex, developmental mechanisms.

3.1.1.   *Generation of naive T lymphocytes*

The fact that the thymus involutes with age raised the questions as to whether the aged thymus continues to generate naive lymphocytes and whether their functional potential is as in the young. In view of the fact that T cells develop from extra-thymic stem cells, it was of interest to determine whether the reduced level of naive T cells in the aged results from aging effects on the stem cell pool.

A variety of studies have led to the conclusion that T cells are continuously generated throughout the lifespan. Firstly, studies on centenarians have demonstrated substantial levels of naive T cells [10, 43]. Secondly, TdT expression in the thymus was shown in aging persons (70 years), despite the subtotal physiological involution [44]. More recently, an innovative assay has been based on the fact that excision circles created during T-cell receptor (TCR) rearrangements (TRECs) are present exclusively in naive T cells that have not yet undergone any cell division. Such studies on human peripheral blood cells showed high levels of TRECs in elderly subjects, suggesting the occurrence of newly generated T cells [45]. The findings that these recent human thymic emigrants respond to costimulatory signals, and the fact that the adult thymic stroma can support thymocyte viability in organ culture provide further evidence that the thymus retains ability to orchestrate T lymphopoiesis even late in life [46].

The naive T cell compartment in aging is a mosaic of potentially active, as well as anergic cells [47]. The mechanisms underlying anergy are not yet adequately understood. Anergy is related, at least in part, to altered patterns of signal transduction, as demonstrated in the lymphocytes of aged mice [48-50]. This could be associated with decreased telomere length in the progenitor cells. Hence, anergic naive cells may represent derivatives of stem cells that have been at a stage close to replicative senescence prior to their differentiation into T cells, and therefore the residual replicative potential of the progeny is limited. Although telomerase is upregulated following antigen stimulation [51], it may be active only temporarily [52] and telomere length may not be adequately reconstituted.

3.1.2.   *Activated/memory type T lymphocytes in aging*

A shift from naive to memory type cell populations was envisaged as a result of long-term exposure to a variety of antigens in the peripheral tissues. Another hypothesis to account for this phenomenon [53] relates to the increased apoptosis of naive (CD45RA+) cells upon activation, thus resulting in a relative increase in proportion of memory type cells (CD45RO+). However, it appears that T cells generated in the old may actually exhibit a priori characteristics of memory type cells [54-56]. Hence, reconstitution of T lymphocytes in aged mice following depletion (e.g., by irradiation) resulted in a cell population profile of naive and activated/memory types as observed in the aged. The newly generated T cells in the aging microenvironment manifested phenotypes of "memory/activated cells" and the mRNA showed cytokine production profiles as in aged mice [55, 56].

The increased values of memory type T cells in the peripheral lymphoid tissues may result from altered patterns of thymocytopoiesis rather than strictly due to post-thymic events. Interestingly, IL-7 that is critical for T cell development in the thymus [57-59], was shown to play a role in induction of CD45RA+ cell expansion, while maintaining this phenotype [59]. Accordingly, reduced availability of IL-7 in the aging thymus may be an additional cause of CD45RA+ cells deficiency.

3.2.   Plasticity and regeneration capacity of the thymus

The thymus is unique in its capacity to regenerate following severe lymphoid depletion. This plasticity is deduced particularly from the seasonal thymic involution in reptiles that is followed by regeneration [12-14], as well as spontaneous recovery from acute stress [60] and sublethal doses of irradiation [61, 62]. Radiation recovery was also shown *in vitro*, in organ cultures of fetal thymus lobes [63]. The fact that this plasticity is not manifested in the age-related involution suggests different, and/or additional mechanisms. Furthermore, the capacity of the thymus to regenerate following severe depletion decreases in advanced age, as revealed from several experimental models.

Noteworthy, irradiation of splenectomized mice resulted in a failure of the thymus to recover [64], suggesting contribution of the spleen to processes of thymic recovery. This could be related, at least in part, to essential cytokines that are provided by the spleen, and/or effects of chemoattractants.

Thymic recovery following irradiation was examined by transplantation of identifiable hematopoietic stem cells. The results showed that the thymus of aged mice retained the ability to attract progenitor cells. Studies on rats showed that the vascular structures were destroyed on day 3 after 6 Gy, but recovered by day 7 [65]. Whereas the cortical epithelial cells contributed to this recovery, the medullary epithelial tissue remained inactive for a relatively long period. A defect was also detectable after transplantation of the cells directly into the thymus [61], suggesting a failure of the aged thymic microenvironment to support T cell development from progenitor cells.

The ability of hematopoietic tissues of aged mice to provide cells that were able to repopulate the thymus of an irradiated young animal seemed to persist throughout the lifetime of the donor. This was inferred from *in vivo* studies on radiation chimeras, as well as from reconstitution of genetically anemic mice [66], and from *in vitro* studies in which the bone marrow cells of aged mice were applied directly onto fetal thymic explants [67]. Competitive colonization assays have proved powerful for more critical evaluation of aging effects [68, 69]. Hence, competitive colonization assays did reveal quantitative and qualitative differences between cells of old, compared to young donors, both *in vivo* [69] and *in vitro* [67, 70]. *In vitro* competitive colonization enabled critical quantitative and qualitative analyses of these age-related changes at the bone marrow level, as discussed in detail (see Section #7).

The relevance of competitive colonization to normal regeneration and continuous developmental processes was inferred from the observation that donor type bone marrow cells competed with intrinsic thymic resident cells [71]. Inferiority of the old donor derived bone marrow cells was also manifested under such conditions.

Another aspect of thymus regeneration relates to the stromal elements. Given the proper conditions, the thymic stroma may have the capacity for regeneration in the absence of extrinsic input of new stem cells. Bone marrow derived stroma cells in the thymus were shown in irradiated recipients that were reconstituted with hematopoietic stem cells from identifiable donor source [72].

Studies attempting at expansion and maintenance of thymic stromal tissue used a semi-organ culture method, in which the bulk of all thymic stromal cells were cultivated in a mixed monolayer cell culture [73]. A key element in this method was that it depended on the prevention of fibroblast growth. These studies revealed that the mixed cell culture could act as an inductive thymic microenvironment in support of T cell development from stem cells. In addition, they indicate the capacity of these cells to expand under suitable *in vitro* conditions. However, thymic stroma that was obtained from aged mice failed to grow under such conditions [Globerson, unpublished]. This observation could be explained on the basis of a failure to replicate due to replicative senescence of the cells, or an absence of critical growth factors, or inhibitory effects. Interestingly, when thymic stroma from aged mice was cultivated without elimination of fibroblast growth, a limited appearance of a variety of thymic stroma cells, including epithelial and dendritic cells, was enabled [Globerson, unpublished]. It thus appears that the potential for regeneration of the thymic stroma is retained to a certain extent in old age, although it is not readily manifested.

Vitamin E enhances T cell differentiation in the thymus through the increase of stromal thymic epithelial cells. The effect is related to increased binding of immature T cells to the epithelial cells, via increased expression of ICAM-1 [74]. A variety of mechanisms may thus be responsible to the limited growth of thymic stromal elements.

3.3.   Cell cycle and programmed cell death

Reduced output of newly generated T cells in aging could be causally related to a decline in cell replication and/or to an increased rate of programmed cell death in the aging thymus. In fact, levels of both mitoses and apoptoses are decreased in the aging thymus.

A decline in cell replication in the thymus has been documented from several laboratories. Studies on cell replication in the thymus were performed both *in vivo* and *in vitro*. *In vivo* studies were based on injection of 3H-Thymidine and subsequent analysis of thymocytes by autoradiography [62]. In addition, *in vitro* studies involving BUDR labeling of mitogen (concanavalin A) stimulated cells in thymic organ cultures, and subsequent flow cytometry [75], reinforced the conclusion that the rate of cell replication in the thymus is decreased in advanced age, also following arbitrary triggering the cells to divide. The results of all of these studies provided further support to the idea that aging is associated with reduced cell replication in the aging thymus.

Studies monitoring cells in apoptosis in the aged thymus disclosed decreased, rather than increased levels. Reduced levels of apoptosis were revealed from flow cytometry analyses of thymic cells, showing a decreased sub-G0 fraction in thymocytes of old, compared to young adult mice. Spontaneous apoptosis following 24 hour incubation of thymocytes *in vitro* was also decreased in thymocytes of the aged mice [Globerson, unpublished].

Studies on p53 deficient mice suggested that accelerated development and aging of the immune system may be related to loss of regulation of cell cycle, DNA repair and apoptosis [76]. It thus appears that regulation of T cell development in the thymus is altered with age.

A variety of mechanisms may account for these age-related phenomena. Particularly, neuro-endocrine control has been extensively examined in relation to aging, as further described (see Section #4).

3.4.   Mature cell dynamics in the thymus: outside-inside cellular traffic

A pure synchronic linear development from stem cells to the diverse cell populations in the absence of any other mature or intermediate cell types is unique to the first wave of early embryonic thymocytopoiesis. Aside from this exceptional situation, the thymus contains a variety of cell types and their coexistence with less mature T cell types, as well as stromal cells, may contribute to regulation and control of ongoing developmental processes. In addition, re-entry of mature T cells from the peripheral tissues into the thymus needs to be considered. The occurrence of such cells has been demonstrated, particularly under conditions of chronic viral infection or antigen stimulation [77-80].

Theoretical modeling of regulation of thymic cellularity suggested that the balance of cell numbers cannot be based solely on intrinsic processes and that one needs to consider also the role of cells entering the thymus from extrinsic sources [81].

In vitro studies using co-cultures of isolated mature and immature T cell populations revealed effects of the mature lymphocytes on T cell development [82]. Regulatory effects of mature CD4 T cells were also shown in T cell development from human umbilical cord blood cells, under similar *in vitro* experimental conditions [83]. Mathematical simulation of cell interactions in the thymus led to the hypothesis that T cell development in the thymus is regulated by the more mature thymocyte subsets [7]. Interestingly, cross-interactions of cells in the thymus were shown between thymocyte subsets and the thymic microenvironment [8, 84]. In addition, induction of thymic cortex is regulated by subpopulations of pro-thymocytes [84]. Peripheral lymphoid cells of aged mice had different regulatory effects on thymocytopoiesis compared to peripheral cells of the young [82]. The presence of CD4 single positive lymphocytes in cultures of CD4/CD8 double negative thymocytes originating in different Thy1 congenic donors respectively, led to a decrease in development of CD4/CD8 double positive and CD8 single positive in the old-donor derived cells, as compared with the young ones. In contrast, CD8 single positive cells had no such effect.

Altered patterns of T cell development, and the decreased output of naive cells in aging, may thus be affected by a variety of mechanisms. These include effects of cells in progressive stages of differentiation on the less mature phenotypes, and on stromal elements, as well as incoming mature T cells, that may have been subject to long term exposure to antigens *in vivo*.

The signal(s) leading mature T cells to exit from the thymic medulla represent an additional important aspect of T cell development. Our own studies pointed to T cell exit from the thymus to the peripheral lymphoid tissues upon *in vivo* antigen stimulation [85]. This was demonstrated in the *in vitro* experimental model for induction of an antibody response to a hapten conjugated to a protein carrier molecule. When splenocytes from mice immunized to the carrier protein were challenged *in vitro* with the hapten conjugated to the specific carrier, they produced anti-hapten specific antibodies. However, thymectomy prior to the *in vivo* immunization interfered with the subsequent splenocyte response *in vitro* to the hapten. It was thus concluded that *in vivo* immunization triggers thymic cells to migrate to the peripheral lymphoid tissues. It is tempting to assume that the cells enter cycle before emigration, in line with the studies of Scollay [86].

Since cells in the aging thymus show lower levels of divisions, this may be an additional cause of reduced output of newly generated cells from the aged thymus. Reduced output of the newly generated cells may thus account, at least in part, for the limited generation of T cells.

3.5.   Alternative developmental processes in the aging thymus

The thymus is not an obligatory site for T lymphocytes and it can also accommodate other cell types. A small number of B and myeloid lineage-restricted progenitor cells were observed in the fetal thymus [87]. Studies on aging revealed that under certain conditions cell types other than T lymphocytes can also develop in the thymus, as further described.

3.5.1. *NK cells in the aged thymus*

Aging is associated with an increase in the proportion of T cells that co-express natural killer (NK) cell markers [88]. We have observed increased values of NK-1+ cells in the aging thymus, as well as in fetal thymus lobes repopulated with bone marrow cells from aged mice [89]. What determines the intrathymic development and lineage branching of NKT cells, given that they are absent in the neonatal period is not yet clear.

3.5.2. *Occurrence of B cells in the aged thymus*

Occurrence of B cells in the aged thymus was reported from several laboratories [90-92]. Studies on the murine AKR leukemia model showed B cell leukemia development in the thymus. This has led to the hypothesis that lack of favorable microenvironment for potential lymphoma cells development in the T cell pathway enables a B cell developmental route [90]. The basis of B cell occurrence in the aging thymus may be similar. Co-cultures of aged bone marrow cells seeded onto fetal thymus explants [67] under conditions favoring T cell development did not show abundance of B cells. The reconstituted thymus explants showed limited B cell function, as measured by the response to LPS stimulation, in contrast to the increased response observed in the intact aged thymus. Hence, normally, B cells can develop from hematopoietic stem cells, but this is not the default pathway of the system in the thymus. It may be based on the properties of the aging stem cells, or the thymic microenvironment, as well as other causes.

Development of myeloid cells in the thymus was observed in cultures of adult irradiated thymus in the presence of bone marrow explants [28]. Increased levels of macrophages were observed in the aged thymus *in vivo* [89], as well as following seeding of bone marrow cells from aged mice onto lymphoid depleted fetal thymus explants.

3.6.   Thymic factors and cytokines in the aging thymus

The thymus itself produces a variety of factors that affect T cells in development. An age-related decrease in production of these factors has been widely documented during several decades, indicating that certain age-associated changes in thymic function and in T cell differentiation may be affected by deficiency in such factors. This area has been reviewed in detail recently [93-95].

A variety of cytokines are produced in the thymus [25, 96], and may play significant roles in the early developmental processes in the thymus. Particular attention is drawn to IL-7, for its critical role in promoting the transition from the CD4/CD8 double negative to double positive phenotypes [57, 58, 97-99], and its production in aging is decreased, as inferred from *in vitro* studies on bone marrow cells [99].

Another cytokine of interest is TNF-alpha, since recent studies suggest that it may promote the appearance of CD4/CD8 double positive cells [100]. Age-related changes in the profile of cytokine production and availability may thus be responsible, at least in part, for the pattern of thymic function in aging. Other factors that need to be considered are the integrins and chemoattractants [102-103].

4. Neuro-endocrinological aspects of thymic involution

4.1. Is thymic involution inevitable?

Age-related involution of the thymus was originally described by Hammar [cf. 104] as a basic histogenetical rule, and it was assumed to represent "Alternsinvolution". The fact that the thymus is the first site where processes of aging are manifested has led to the notion that the thymus plays a role of the "pace-maker of aging". However, a variety of studies indicated that the thymus is actually subject to hormonal control, as part of the Neuro-endocrinological axis, and receptors to a variety of hormones and neuropeptides have been shown in thymic cells.

Involution of the male thymus following puberty depends on the increasing values of testosterone, and the changes in the thymus are manifested mainly in the decrease of thymic lymphoid-cell elements. Castration prevents thymic involution [105]. Tamoxifen treatment reverses thymic involution in intact adult male rats in a dose-dependent manner [106]. On the other hand, tamoxifen administration at pharmacological doses to adult castrated rats results in thymic regression. It thus appears that tamoxifen can reverse thymic involution by reducing testosterone levels, whereas in the absence of testosterone it has thymolytic effects.

Hormonal roles in thymic involution were further indicated from studies in which treatment with growth hormone, as well as prolactin or IGF-1 [107-114] resulted in increased cellularity of the thymus. Treatment had no effect on the nature of the developing T cells, and did not lead to "rejuvenation" of the thymus, suggesting that thymic involution is caused by multiple mechanisms. Studies based on *in vivo* and *in vitro* treatments demonstrated that the effect was exerted on cells in the bone marrow, prior to their migration to the thymus, or at an early stage of their development in the thymus [107]. Hence, at least part of the manifestations of thymic involution are associated with a decreased migration of stem cells and their homing to the thymus.

4.2. Expression and function of sex-steroidal hormones in the thymus

Receptors to sex-steroidal hormones are expressed in the thymus, particularly in reticuloendothelial cells [115, 116] yet also in early immature lymphoid cells [117, 118]. We have demonstrated the expression of estrogen receptors in murine thymocytes of both females and males, and there was no indication to an age-related decrease in levels of estrogen-receptor expressing cells. However, only thymocytes of the females reacted to *in vivo* treatment with estradiol (E2) as manifested in an increase in creatine kinase activity, while the males reacted to testosterone [119]. Interestingly, cell replication was augmented by *in vitro* treatment with E2 to FTOC colonized with bone marrow cells or thymocytes from young adult (2 month) and not aged (24 months) mice. Hence, whereas ER is functional in the aged mice, its triggering does not result in cell replication.

More recent studies showed that estrogen alpha receptor (ERalpha) is necessary in thymic development [116]. The study was carried out on ERalpha knockout mice, that have significantly smaller thymuses compared to the wild type littermates. By establishing bone marrow radiation chimeras between the KO and the WT mice it was found that the small thymus was due to lack of ER in the radiation - resistant tissues rather than the hematopoietic cells. However, the KO mice did respond to treatment with estradiol, and showed thymic atrophy as a result of that treatment. Accordingly, the estradiol-induced thymic atrophy is mediated by another receptor pathway. These observations are in line with our findings of ER in thymic stromal cells, as well as lymphocytes [118].

These observations lend further support to the hypothesis that hormonal effects leading to increased cellularity in the thymus cause stem cell migration and their possible expansion, rather than thymocyte cell divisions.

Glucocorticoids are known to cause thymocyte apoptosis at high concentrations. Interestingly, transgenic mice in which the expression of the glucocorticoid receptor was down-modulated specifically in thymocytes showed abnormal thymocyte differentiation, suggesting a significant role of the hormone in T cell development [119]. Studies on T cell development in organ cultures of fetal thymus lobes from transgenic mice, following inhibition of glucocorticoid biosynthesis led to the conclusion that the hormonal effect in the thymus may be related to its local synthesis in the thymic epithelium.

4.3.   Neurotransmitters in the aging thymus

Direct effects of neuropeptides and neurotransmitters on thymic epithelial cell function, were shown by treating rat thymic epithelial cells in culture with various agents [120]. Cell proliferation increased following treatment with CGRP and histamine, but decreased by isoproterenol, 5-HT and acetylcholine. Thymocytes themselves are the source of acetylcholine that may play a role in the dialogue between thymocytes and thymic stroma [121, 122].

Neuropeptide control of thymocytopoiesis was inferred from studies on apoptosis in thymic organ cultures [123]. The data showed that the effect was due to thymic epithelial cells. Fetal thymus explants responded to cholinergic stimuli in increased apoptosis, when co-cultured with cortical thymic epithelial cell lines. This was not detectable when the epithelial cell lines were from the thymic medulla. Restructuring of the thymic microenvironment in aging may thus involve changes in such stromal cell components, and/or in the availability of the relevant neuropeptides [124]. Cholinergic stimulation of thymic epithelial cells *in vivo* could be provided by innervation [120, 124] and/or by thymic lymphocytes [122].

Although these experiments indicate a response to cholinergic stimulation, the possibility of adrenergic effects under physiological conditions *in vivo* should also be considered [121]. It should be noted in that respect that the density of noradrenergic sympathetic nerves and the concentration of noradrenalin increase dramatically in the thymus of old mice and rats [125].

5.   Hematopoiesis in aging

5.1.   The lymphohematopoietic stem cell compartment in aging

Basal hematopoiesis is maintained in normal elderly humans [6, 126, 127], and in healthy centenarians [10, 11]. Although the proportion of CD34 positive cells does not change with age, a limit in the capacity for hematopoiesis becomes apparent under conditions of hematological stress [6]. Assessment of the developmental potential of the stem cells has therefore clinical significance. This issue is also important for autologous transplantation of stem cells in hematopoietic compromised patients. The recent report by Siegel et al. [128] showed that mobilized stem cells from multiple Myeloma patients were as effective as those from young patients for transplantation in course of clinical treatment. However, the question of whether such transplants are functional for long-term will need to be answered.

The mechanisms underlying the decline in ability to cope with hematological stress and to upregulate hematopoiesis under such conditions [6] may be complex and have not yet been fully elucidated. One possibility is that there is a shift in the cytokine network. Indeed, studies on cytokine production by bone marrow cells revealed a shift from Th1 to Th2 types of cytokines [M. Dayan, E. Mozes and A. Globerson, unpublished], similarly to observations on the spleen [129, 130]. Furthermore, Buchanan et al. [131] demonstrated that light density leukocytes from aged subjects released less colony stimulating factor (CSF) than cells from the young and GM-CSF accounted for 72-100% of CSA of the young and only 0-42% in case of the old. This idea was further substantiated by the findings on the expression of mRNA for GM-CSF. Similar observations were made on mice. Interestingly, GnRH agonist treatment was found to decrease the levels of Sca-1 expressing cells in the bone marrow and to increase their levels in the thymus of C57BL/6, but not BALB/c mice [114]. Hence, mechanisms enhancing mobilization of stem cells to the thymus may be different under different genetic backgrounds. Manifestation of genetic differences in patterns of aging effects on hematopoiesis in different mouse strains [133, 134] should be considered in extrapolation from observations on mice to humans.

5.2.   Status of the stem cell compartment in aging

Stem cells are characterized by their potential to give rise to the different types of blood cells and by their capacity for self-renewal. The continuous generation of functional heterogeneity among the clonal progeny of hematopoietic stem cells is in support of intrinsic control of stem cell fate and provides a model for long-term maintenance of hematopoiesis [135]. Proliferation induced decline of primitive hematopoietic cell activity has been suggested, showing that it is coupled with an increase in apoptosis of *ex vivo* expanded CD34+ cells [136].

The questions of whether hematopoietic stem cells maintain their developmental potential throughout the lifespan and whether the stem cell pool becomes exhausted with age, have raised considerable controversy, as reviewed [6]. Recent studies have drawn attention to the fact that tissue committed stem cells are not restricted to the extent as originally thought [137] and that they may remain pluri-potent, capable of giving rise to diverse cell lineages (e.g., neuronal, endothelial, etc.). This point raises the additional question of whether the functional potential of hematopoietic stem cells is affected by age.

Studies designed to examine the hematopoietic potential of the stem cell pool were based on transplantation of bone marrow cells to hematologically compromised mice (irradiated, or genetically anemic mice). The results indicated efficient hematological reconstitution, as exhibited by long-term erythropoiesis [61]. Interestingly, T lymphocyte values decreased eventually in such hematologically reconstituted mice [138, 139], suggesting a selective decline in the capacity to generate T lymphocytes. Accordingly, it is necessary to distinguish between normal hematopoiesis *in vivo* as related to the chronological age of the individual, the intrinsic potential of stem cells for self-renewal, and the capacity to generate the different blood cell types.

It should be noted that the classical studies have used bioassays involving transplantation of bone marrow cells and the data was extrapolated to the function of stem cells within the bone marrow inoculum. Such studies suggested that aging is associated with a relative increase in the proportion of the more primitive stem cell types [140].

Establishment of tools to identify stem cells by virtue of distinct membrane markers has led to major progress in this area. Methodologies that are currently available have made stem cells more accessible for transplantation, for gene therapy and for further analysis. Innovative techniques enable stem cell isolation and extensive expansion *in vitro*, as well as mobilization of hematopoietic stem cells to the peripheral blood, *in vivo*. Stem cell senescence can now be approached both in terms of intrinsic potential of the cells for self-renewal and their hematopoietic function *in vivo*. Indeed, studies on isolated hematopoietic stem cells from aged mice were carried out more recently [141].

Characterization of hematopoietic stem cells in mice is based on distinct phenotypes, that enable identifying early pluripotent cells and more advanced committed cells. Our studies on mice [141] showed a five-fold increase in the proportion of cells bearing the stem cell phenotype Thy1.1^{low}Sca-1$^+$Lin$^-$, and a decrease in the multipotent progenitor subpopulations (Mac-1^{low}CD4$^-$c-kit$^+$, Mac-1^{low}CD4$^-$, Mac-1^{low}CD4^{low}). Cells expressing the stem cell phenotype Thy1.1^{low}Sca-1$^+$Lin$^-$ harvested from either young or old mice could reconstitute irradiated mice to the same extent. However, transplantation under limit-dilution conditions disclosed a higher frequency (five fold increase) of the cells in the bone marrow of old mice compared to the young, but they were only one-quarter as efficient at homing to and engrafting the bone marrow of irradiated recipients.

Aging mice showed a relative increase in Sca-1 positive cells that are in cycle (G2/M), in contrast to the situation in the young, where these cells are mostly quiescent [141]. These observations conform to our earlier findings on bone marrow cells cultured on bone marrow derived stroma cells [142], as well as upon seeding onto fetal thymus explants [143].

Since entering cycle is induced by cytokines, the fact that stem cells are a priori in cycle may be due to age-related altered profiles of cytokine production, as reviewed [42] and further elaborated more recently [130].

It should be emphasized that although hematopoiesis is reconstituted following stem cell transplantation, generation of T cells under such conditions is hard to achieve. The basis of this problem is as yet unresolved. Several hypotheses can be proposed to account for this phenomenon.

(1)   The transplanted cells may lack the expression of homing markers, to enable their migration and settling in the thymus. Stem cells from aged mice behaved similarly to mobilized stem cells from young adult donors [141, 144].

(2)   Hierarchy in the types of blood cells that develop in ontogeny as well as in regeneration from sublethal doses of irradiation and following bone marrow transplantation [145] shows initially erythrocytes and granulocytes, and then B lymphocytes. The T cells are the last ones to appear under all of these conditions. This suggests control mechanisms that affect stem cell migration to the thymus, and/or sequential expression of growth factors and cytokines that control the development of the different cell types within the bone marrow.

5.3.   Stem cell migration and homing

The observation that HSC from aged mice behave similarly to mobilized hematopoietic stem cells [141] suggests that a better understanding of aging effects on stem cell migration and homing is required.

The chemokine stromal derived factor-1 (SDF-1) and its receptor CXCR4 were found critical for the engraftment of human stem cells in the bone marrow of SCID mice [146, 147]. It was thus suggested that migration to SDF-1 is associated with localization of stem cells in the bone marrow, and that differentiating cells with reduced migration levels are prone to exit into the blood circulation.

The relevance of SDF-1 to stem cell migration was indicated from an additional angle. Murine hematopoietic progenitor cells were shown to roll along bone marrow microvessels that display selectins and integrins, while SDF-1 is expressed by human bone marrow vascular endothelium [147]. *In vitro*, human CD34+ cells establish efficient rolling on P-selectin, E-selectin and the CD44 ligand hyaluronic acid, under physiological shear-flow conditions. In the presence of surface bound SDF-1, CD34+ progenitors rolling on endothelial selectin rapidly developed firm adhesion to the endothelial surface, mediated by an interaction between ICAM-1 and its integrin ligand, which co-mobilized with SDF-1. The possibility that aged cells are unable to down-regulate CXCR4 expression in the bone marrow thus deserves attention. High levels of CD44+ cells in the bone marrow and thymus of aged mice [148], suggested that down-regulation of CD44 expression is less efficient in aging. It would appear that the cells remain adherent to stroma elements and fail to proceed in subsequent steps of differentiation. Mathematical modeling of the dynamics of cell adherence and detaching from the stroma elements [149, 150] raise the possibility that such mechanisms account for the generation of T cells in the aged thymus.

It may be speculated that hormonal effects on mobilization of stem cells from the bone marrow to the thymus as described above (see Section #4.4) may be based on down-regulation of CD44, and/or CXCR4. However, the effects of aging on expression of CXCR4 receptor and its SDF-1 ligand in the bone marrow are as yet unknown.

Other chemokines that play a role in the adherence of thymocytes (e.g., TECK) may be relevant in this respect [151, 152], yet their status in aging is still unknown.

6.   Do lymphohematopoietic stem cells age?

The question of whether stem cells undergo senescence has been approached by both *in vivo* and *in vitro* methodologies. The different experimental strategies led to similar conclusions, as further described.

6.1.   Studies based on *in vivo* experimental models

The classical studies were based on serial transplantation of limited numbers of bone marrow cells to irradiated mice and monitoring the number of hematopoietic colonies that developed in the spleen (CFU-S). These studies revealed a gradual reduction in the number of CFU-S that developed after 4-5 serial transplantations, and eventually there were no detectable colonies in the spleens of the recipient mice [5, 153]. These findings suggested that the hematopoietic stem cells have a limited capacity for self-renewal. However, the relevance of this finding to aging was questioned, in view of the fact that basal hematopoiesis is maintained through old age. It should be noted that these studies were based on the assumption that transplanted bone marrow cells home to the bone marrow of the recipients. Accordingly, it was taken for granted that the bone marrow of the recipients throughout the serial transplantations would contain descendent stem cells from the original donors. Yet, as we know today, the stem cells in the bone marrow transplanted serially to irradiated mice may have actually lost their homing capacity.

Another approach to that question was based on serial hematological depletion and subsequent spontaneous recovery (e.g., by treatment with hydroxyurea [154], or irradiation [155]) within the individual mouse. The results led to the conclusion that enforced hematopoiesis within the individual mouse does not lead to a decline in stem cell function.

These *in vivo* studies are hard to interpret in relation to aging, because the experiments were performed on young mice. The total stem cell compartment in the young individual mouse may have been capable of coping with the induced hematological stress under the experimental conditions. The question of whether the stem cell pool is affected by aging thus remained unresolved.

6.2.   *In vitro* studies

More recent studies have been based on *in vitro* cultures of isolated hematopoietic stem cells. *In vitro* expansion of hematopoietic stem cells (CD34+CD38-) led to the claim that the cells can be expanded *in vitro* indefinitely, and thus provide a powerful tool for autotransplantation for a wide range of possible treatments, in addition to hematological reconstitution. Search for methods that will enable a quantitative expansion of HSC has been a goal for both basic research and clinical purposes. To that end, a variety of stroma cell lines have been developed to support stem cell proliferation. Our studies

[142] represent the particular focus on hematopoietic stem cells in aging. Expansion of murine stem cells was examined in cultures in the presence of stroma cells (14F1.1) that selectively support stem cell long-term maintenance and expansion. Interestingly, bone marrow cells from aged mice showed initially higher proliferation rates than the young. In addition, they also generated increased levels of myeloid cells (CFU-C and CFU-S), in line with the observations on elderly human subjects [156], as well as our more recent studies described above.

Does replicative senescence result in narrowing the range of developmental potential of the cells?

Cells that expanded on the stroma cell monolayer could also reconstitute FTOC and give rise to T cells over a period of several weeks. However, manifestation of aging became apparent when a competitive colonization assay was used. Competition assay was based on the following assumption. If stem cells of the old are not different than those of the young, then transplanting them in a mixture to the same individual recipient will result in progeny blood cells derived from both donors at the same proportions as in the transplanted inoculum [69]. This strategy was applied to the *in vitro* experimental model. After one week of culture on the stroma cell line, both young and old-donor derived cells failed to compete with fresh bone marrow cells in reconstituting the FTOC, although they could develop under these conditions in the absence of competition [142]. Stem cells of the young thus seemed to have aged *in vitro*, in the sense that they behaved as fresh bone marrow cells from aged donors under similar conditions.

More advanced methods were established for maintenance of HSC without the need for stroma cell support. This was enabled by using a combination of relevant growth factors, particularly, Stem Cell Factor (SCF; c-kit ligand), FLT3 ligand, megakaryocyte growth and development factor (MGDF) and IL-6 [157]. The *in vitro* propagated cells showed maintenance of hematopoietic function, manifested in reconstitution of non-obese NOD-SCID mice.

A critical examination is still due to determine whether the *in vitro* expanded cells preserve their properties indefinitely, or whether they eventually age. So far, studies have focused on the capacity of the cells for hemopoiesis, with no data on the generation of T cells. Furthermore, the question as to whether the expanded stem cells maintain their capacity to express the entire spectrum of developmental potential as that of freshly drawn stem cells from young donors, is still open.

6.3.   Telomerase function in stem cells

Telomerase expression has been correlated with replicative capacity of somatic cells and with the residual potential for replications. It has been proposed that the sequential loss of telomeric DNA from the ends of the chromosome with each somatic cell division eventually reach a critical phase when replications stop, and the cell senesce. Telomerase is expressed in germline cells, where telomere length is maintained and intact chromosomes can be transmitted to the subsequent generation. It could be predicted that stem cells maintain their replicative potential, and that they express telomerase constitutively to ensure long-term fidelity of self-renewal. Whether stem cell replication is associated with a decrease in telomere length has thus been a question of major interest.

Vaziri et al. [158] demonstrated that CD34⁺CD38^low cells purified from human bone marrow have shorter telomeres than cells from fetal liver or cord blood. Sakabe et al.

[159, 160] demonstrated that human cord blood-derived CD34⁺CD38⁻, as well as CD34⁺c-kit⁻ cells, characterized as primitive progenitors, have low telomerase activity.

Low levels of telomerase in primitive hematopoietic cells (CD34⁺CD71^lowCD45RA^low), and higher levels in the early progenitor cells (CD34⁺CD71⁺) were also shown by Chiu et al. [159]. Non-expanding CD34⁺ cells had low, or undetectable telomerase levels. Accordingly, there is a distinction between primitive and more advanced stem cells, in terms of telomerase function. Telomerase activity is present at basal levels in primitive stem cells that are cycle-quiescent, and is upregulated by cytokines. This point is important for long-term maintenance of functional stem cells *in vitro*, in the presence of cytokines. It should be noted, though, that telomerase function may reduce, but not necessarily prevent, telomere shortening in course of replication [160].

A decrease in telomere length *in vivo* has been noted following bone marrow transplantation [161]. Telomeres in the donor-derived peripheral blood cells of the recipients were shorter than in the respective donors. It thus appears that telomere shortening resulted from enforced extensive replication of the transplanted cells [162].

7.   Generation of T lymphocytes from aging stem cells

The approach to examine whether aging of haematopoietic stem cells may have any impact on a selective decrease in generation of T lymphocytes is of particular interest. To focus on aging of the bone marrow as separate from the issue of thymus involution, we seeded bone marrow cells from young or old mice onto FTOC under conditions favoring T cell development [67, 70]. The number of lymphocytes harvested from the cultures under these conditions did not disclose any obvious difference between young and old donor-derived cells. However, a difference became apparent under competitive reconstitution, when identifiable cells of the young and the old donors were seeded onto the same individual thymic lobe [67, 70, 143]. Identification of T cells that derived from both donors was possible by using congenic mice that express different Thy.1 allotypes (Thy1.1 and Thy1.2).

Competitive-reconstitution assay is thus a powerful tool to reveal manifestations of aging in bone marrow derived cells. Based on these observations, we designed further experiments to assess aging effects on bone marrow cells that might have an impact on T cell development. Interestingly, although the frequency of bone marrow cells that colonized the thymic lobes was reduced with donor age, it had no impact on the total cell number by the first week of culture. This was related to the fact that bone marrow cells of old donors replicated upon seeding in the thymus, while cells of the young started replication only 24 hours later [143]. However, cells of the old showed a decreased capacity for subsequent sequential cell replications [70]. The data suggest that the stem cells in the aged are a priori in cycle, yet their potential for subsequent replications is limited.

Cells of the old gave rise to increased levels of myeloid cells [83], in line with observations on elderly human subjects [156]. More recent studies, using isolated hematopoietic stem cells from aged mice showed a similar tendency for increased myelopoiesis [141]. The aged bone marrow derived cells in the FTOC showed also increased levels of cells expressing the stem cell marker c-kit [83] and the adhesion molecule CD44 [148], suggesting a limit in down-regulation of these markers.

In addition, bone marrow cells of aged mice failed to down-regulate the expression of two distinct membrane polypeptides (pI 5.25 and 5.2; Mr 56 × 10³ respectively) in the thymic stroma upon settling in the thymic stroma under these conditions [163].

Recent studies on human cells, based on the FTOC experimental model, showed that bone marrow CD34 cells from old donors generate *in vitro* less T cells than those of the young [9, 164, 165]. The study led to the conclusion that loss of T cell generation capacity was not dependent on thymic involution and it correlated with the chemotherapy treatment.

Analysis of cell membrane poplypeptides was performed on FT stroma colonized with bone marrow cells from young and old donors. Hence, expression of two distinct membrane polypeptides was downregulated soon after young bone marrow cells had settled in the thymic stroma (within 3 days of culture). In contrast, cells from the old donors showed a decline in their capacity to downregulate this expression [163]. Accordingly, age-related effects are manifested in the bone marrow cells at an early phase of interaction with the thymic stroma.

These studies have thus revealed aging effects on T cell development that are related to the bone marrow stem cell compartment, as separate from the issue concerning the effects of thymic involution on generation of T cells in aging.

8.   Extrathymic T cell development

A variety of studies point to extrathymic T cell development [166, 167]. *In vitro* experimental models, in which T cells develop in the presence of distinct, identified stroma cell lines, or a mixture of cytokines lent support to the idea that T cells can be generated in the absence of an intact thymus. The *in vitro* models have the advantage of enabling a critical analysis of the essential requirements for such development. There are many studies on stroma cell lines that support T cell development in the absence of an intact thymus. Hence, bone marrow derived stroma cells were found most efficient in selecting immature thymocytes and promoting their subsequent differentiation [168].

Auerbach et al. [169] found that yolk sac cells from 9 day mouse embryos could differentiate on lines of endothelial cell clones that were established from 12 day embryonic yolk sacs. One of these clones supported expansion of yolk sac hematopoietic cells that subsequently could generate B cells, T cells and myeloid cells both *in vitro* and *in vivo*.

There are also studies on T cell development without any stroma cell support. Hence, Hirayama et al. [170] showed that in the presence of SF, IL-11 and IL-7, primitive lymphohematopoietic progenitors can differentiate to T cell lineages. Pawelec et al. [171] demonstrated T cell development from human umbilical cord blood derived CD34+ cells in cell cultures without support of stromal cells. Their cultures consisted of irradiated allogeneic mononuclear blood cells that provided both antigen stimulation and accessory cells.

Although these are highly artificial experimental conditions, they indicate possible alternative pathways of stem cell differentiation under conditions of thymic deficiency. It is therefore of interest to find out whether T cells are generated in aging at extra-thymic sites. It should be pointed out that the aged bone marrow contains increased (two-fold) amounts of mature T cells, compared to the young [172]. Whereas they may have circulated from the peripheral tissues, the possibility that they represent intrinsic development in the bone marrow cannot be excluded. Accordingly, newly-generated naive T cells detected in the peripheral blood of elderly subjects [10] may represent, at least in part, development in extra-thymic sites.

9.   A "fountain of youth" for the aging thymus?

As shown above, the thymus is dependent on external source of stem cells, at least for the lymphoid cell types, and it is also dependent on extrinsic hormonal control. In addition, there are intrinsic factors that affect lymphoid development, and possibly maintain the anatomical structure of the epithelial cell components. This view of the thymus raises the questions of whether thymic involution is inevitable, and whether an involuted thymus can regenerate. A close examination of checkpoints in thymocytopoiesis revealed a decline in the capacity for cell replication, as discussed above. This feature in itself seems to be based on the stem cells properties, probably related to reduced telomere length [6].

Another critical checkpoint is at the transition of CD4/CD8 double negative cells from the phase of CD44+CD25- to CD44-CD25+. This stage is also characterized by the onset of expression of RAG. The aging thymus is characterized by a decreased capacity to downregulate CD44 and to express CD25 [54], as well as RAG-1 and RAG-2 [173, 174]. It thus appeared that the decline in these critical processes is fundamental in leading to thymic involution. Indeed, studies on transgenic mice lent support to this hypothesis [174].

The question of whether the decrease in RAG expression in the aged thymus is based on stem cells properties or on the thymic microenvironment led us to examine it under *in vitro* experimental conditions [173]. Hence, fetal thymus explants were co-cultured with either bone marrow cells depleted of mature T cells, or double negative CD4/CD8 thymocytes from young, as well as old donors. The cultures were incubated under conditions enabling T cell development, and were examined at different time intervals, for the expression of RAG. The results showed RAG expression in cells that derived from the bone marrow of both young and old donors, and from the thymocytes of the young - but not those of the old. Accordingly, the decrease in RAG expression is not due to aging effects on the bone marrow cells, but rather on the thymic microenvironment. Furthermore, the mere residence of DN cells in the aged thymus turned them incapable of RAG expression. Whereas the basis of this negative control is still not clear, the results point to manifestation of aging effects on the thymic stroma.

These effects can be based on structural, quantitative and/or qualitative changes in stromal cell populations, as suggested from morphometric analyses of aging thymuses [33]. The effects may also be related to limited availability of distinct cytokines, such as IL-7 that affects the transition from CD44+CD25- to CD44-CD25+ phenotypes. Since IL-7 is a non-redundant cytokine in thymic T cell development, and is produced by stromal cells, it is reasonable to consider a limit in a distinct stromal cell type in the aging thymus. It is of interest to note in that respect that the cortical components of aged thymuses, compared to the young, do not express the 6C3 marker [175, 176], as revealed from analyses based on immuno-histochemistry, immuno-fluorescence and flowcytometry [Globerson, unpublished]. Whether cells expressing this marker develop intrinsically in the thymus, or whether they derive from an extrinsic source of stem cells, is as yet unknown.

In view of the observed effects of growth hormone and IGF-1 treatment, it seemed that different mechanisms account for the mere increased cellularity. These may include mobilization of stem cells from the bone marrow, cell replication and apoptosis, whereas expression of RAG and distinct cell phenotypes, and subsequent appearance of the single positive CD4/CD8 T cells expressing the T cell receptor repertoire are not affected. The key question is still whether thymic involution is based on one critical, central cause, that leads to a cascade of events, or whether a combination of different independent mechanisms leads to the same ultimate result.

Can dietary restriction prevent thymic involution? Dietary restricted mice showed higher levels of T cell frequencies in their peripheral tissues, compared to controls [177]. Weindruch et al. [178] demonstrated that in mice under dietary restriction regimens, thymic size did not decrease to the same extent as in controls. However, our own studies using long-lived transgenic mice (a-MUPA) that are genetically dietary-restricted, had involuted thymuses in old age similarly to their age-matched controls [179]. Hence, the improved thymic size observed in the case of enforced dietary restriction may not be related to the dietary restriction per se, but rather to other associated mechanisms. This finding conforms to the idea proposed by George and Ritter [180] that scaling down of the thymus with age has actually been evolutionary advantageous.

10.   Concluding remarks

Recent advancement in understanding of stem sell biology and thymocytopoiesis points at a series of cell properties and processes that may change with age, suggesting multifactorial mechanisms, that vary in their manifestation in different individuals. An outline of selected age-related changes is presented in Table 1.

The following conclusions can thus be drawn from the data gained so far.
- The available data emphasize the need to distinguish between age-related changes in the stem cell compartment within the individual, and possible manifestations of aging of the stem cells, as well as the dynamics of cellular and molecular events along the stem cell-thymus axis.

Table 1. Selected age-related changes and potential manifestations of aging along the stem cell-thymus axis and early processes of T cell development.

Compartment/process	Developmental step	Manifestation	Reference
Bone marrow, stem cells	Hematopoiesis in old age	Maintained	6
	Recovery following hematological stress	Reduced	6
	Frequency of cells	Increased frequency of stem cell phenotypes (x5), yet only about 20% are functional	141
		Increased levels of primitive stem cells	140
	Cell cycle	Increased proportion of stem cells in cycle (G2/M)	141, 142
	Self-renewal	Replicative senescence following enforced chronic replication	5, 136, 153, 155
		Telomere shortening	158, 161, 162
	Cytokines	Altered profile	131, Mozes et al. unpublished
Bone marrow, T cells	Occurrence of mature T cells	Increased numbers of T cells	172
Thymus, lymphoid compartment	Homing and attachment of bone marrow cells to the thymic stroma	Increased levels of CD44+ cells	54, 148
		Decreased affinity to the thymic stroma. Reduced capacity for down-regulation of membrane markers in the stroma	67, 70 163
	Cell replication in the thymus	Reduced	61, 62, 118
	Programmed cell death in the thymus	Decreased levels of apoptosis	118, Globerson, unpublished
	Transition from DN to DP cells	Reduced	54, 148
	Transition from CD44+CD25-to CD44-CD25+	Reduced	54, 148, 174
	Expression of RAG-1 and RAG-2	Reduced	173, 174
	Expression of estrogen receptors in lymphocytes	No change in proportion of cells expressing the receptor; reduced response to estradiol	118
Thymus, microenvironment	Stromal cells (epithelial, dendritic, etc.)	Morphological changes, change in profile of cell membrane marker expression	33
	Stromal cell replication (self-renewal)	Reduced	Globerson, unpublished
	Expression of estrogen receptors	No change	118
	Function of estrogen receptors	Reduced response to hormone, manifested in creatine kinase production	118
	Thymic hormones	Reduced production and function	93

Table 1. Continued.

Compartment/process	Developmental step	Manifestation	Reference
	Cytokines	Reduced IL-7	99
	Neuroendocrine function	Altered neuronal architecture	125
		Reduced cholinergic and possibly adrenergic control	121-125
Thymus, involution	Prevention or reversibility of involution	Castration	105
		Increased thymic size in response to treatment with growth hormone (GH), prolactin, IGF-1	107, 113, 114
		Prevention of involution in RAG transgenic mice	174
		Beneficial effects of vitamin E	74
		Dietary restriction: increased frequency of T cells	177
		Increased thymic size under dietary restriction regimens	178
		No change in dietary-restricted transgenic mice	179
	Alternative developmental pathways	Activated T cell types	54, 55
		NK cells	88, 89
		B cells	67, 87, 90, 91
Extra-thymic T cell development	Bone marrow	T cells can develop in the bone marrow, but there is no direct evidence related to aging	166, 167
	Gut	T cells can develop in gut associated lymphoid tissues, but there is no direct evidence related to aging	40
Peripheral T cells	Naive T cell types	Continued appearance of newly generated cells	10, 44-46

- A comprehensive consideration of stem cells in aging requires critical evaluation of their intrinsic properties and developmental potential, as well as of microenvironmental conditions, including the profile of cytokines, chemoattractants, growth factors, hormones and neuropeptides,.
- Mechanisms that may compensate for aging effects at the tissue, or cell population level, have been demonstrated. *In vitro* experimental models enable elucidation of such mechanisms. Understanding the basis of compensatory mechanisms is a key element in designing strategies to prevent, or minimize the hazards of aging effects.
- It should be noted that although what cells can do under experimental conditions may not necessarily reflect what they normally do *in vivo*, developmental pathways that are unconventional in the young adult may become dominant as alternative pathways in aging.
- Further molecular genetic approaches will be needed to establish mechanisms underlying the differential decrease in the potential for T cell development in aging, to pave the way for relevant intervention strategies.
- Studies of hematopoietic stem cells-thymus axis in aging have important implications beyond the direct relevance to old age. The results of such studies have important implications to hematopoietic stem cell transplantation and designing strategies to ensure efficient T cell development under such conditions, as well as *in vitro* expansion of such stem cells for transplantation and gene therapy.

11.   Conclusion

Thymocytopoiesis starts from stem cells that migrate from hematopoietic tissues, settle in the thymus and differentiate there into T lymphocytes. Generation of T cells continues through old age, albeit at a reduced rate and altered patterns. Whereas these age-associated phenomena have been causally related to thymic involution, contribution of changes in hematopoietic stem cells has also been demonstrated and extrathymic T cell development needs to be considered. Elucidating the fate and functional potential of these stem cells in the old organism requires a comprehensive approach that takes in account cells giving rise to T lymphocytes as well as those developing into the supporting milieu of the thymic microenvironment. Re-evaluation of aging processes is now feasible in the light of current knowledge on stem cell biology that has progressed considerably, due to the growing interest in stem cells for transplantation and gene therapy. Innovative methods enable distinguishing between the status of the stem cell compartment in aging, and the possible occurrence of individual senescent stem cells. This distinction is crucial for understanding the developmental potential of the stem cell pool in the aged. It now appears that aging effects along the stem cell - thymus axis are multi-factorial. The effects include common cell aging processes (e.g., cell replication and programmed cell death), as well as processes unique to the immune system (e.g., generation of the T cell receptor repertoire). Understanding the mechanisms underlying the processes along this axis is crucial for the establishment of novel approaches to maintain a beneficial immune system in old age, as well as for promotion of T cell generation following stem cell transplantation. *In vitro* expansion of long-term functioning stem cells for clinical applications, including gene therapy, is thus one of the arms of stem cell aging biology.

12.   References

1. Morrison, S.J., Uchida, N. and Weissman, I.L. (1995) Annu. Rev. Cell Dev. Biol. 11, 35-71.
2. Auerbach, R. and Huang, H. (1996) Stem Cells 14, 269-280.
3. Dieterlen-Lievre, F., Godin, I. and Pardanaud, L. (1997) Int. Arch. Allergy Immunol. 112, 3-8.
4. Morrison, S.J., Wright, D.E., Cheshier, S.H. and Weissman, I.L. (1997) Curr. Opin. Immunol. 9, 216-221.
5. Globerson, A. (1997) Arch. Gerontol. Geriat. 24, 141-155.
6. Globerson, A. (1999) Exptl. Gerontol. 34, 137-146.
7. Mehr, R., Perelson, A.S. Fridkis-Hareli, M. and Globerson, A. (1997) Immunol. Today 18:581-585.
8. Van Ewijk, W., Hollander, G., Terhorst, C. and Wang, B. (2000) Development 127, 1583-1591.
9. Offner, F., Kerre, T., Smedt, M. and Plum, J. (1999) Br. J. Haematol. 104, 801-808.
10. Bagnara, G.P., Bonsi, L., Strippoli, P., Bonifazi, F., Tonelli, R., DwAddato, S., Paganelli, R., Scala, E., Fagiolo, U., Monti, D., Cossarizza, A., Bonafe, M. and Franceschi, C. (2000) J. Gerontol. Biol. Sci. 55A, B61-B66.
11. Globerson, A. (2000) J. Gerontol. Biol. Sci. 55A, B69-B70.
12. Leceta, J., Garrido, E., Torroba, M. and Zapata, A.G. (1989) Cell Tissue Res. 256, 213-219.
13. El Masri, M., Saad, A.H., Mansour, M.H. and Badir, N. (1995) Immunobiol. 193, 15-41.
14. Alvarez, F., Razquin, B.E., Villena, A.J. and Zapata, A.G. (1998) Vet. Immunol. Immunopathol. 64, 267-278.
15. Moore, M.A. (1997) Stem Cells 15 Suppl. 1, 239-248.
16. Le Douarin, N.M., Dieterlen-Lievre, F. and Oliver, P.D. (1984) Am. J. Anat. 170, 261-299.
17. Auerbach, R. (1961) Dev. Biol. 3, 336-354.
18. Moore, M.A.S. and Owen, J.J.T. (1967) J. Exp. Med. 126, 715-726.
19. Owen, J.J.T. and Jenkinson, E.J. (1981) Prog. Allergy 29, 1-34.
20. Auerbach, R., Globerson, A. and Umiel, T. (1982) in: N. Cohen, and M. Siegel (Eds.), Ontogeny and Phylogeny of the RES. Plenum Press, pp. 687-711.
21. Hofman, F. and Globerson, A. (1973) Eur. J. Immunol. 3, 179-181.
22. Dahl, C.A. (1980) J. Immunol. 125, 1924-1927.
23. Globerson, A., Woods, V., Abel, L., Morrissey, L., Cairns, J.S., Kukulansky, T., Kubai, L. and Auerbach, R. (1987) Differentiation 36, 185-193.
24. Liu, C.P., Globerson, A. and Auerbach, R. (1993) Thymus 21, 221-33.
25. Medvinsky, A.L., Samoylina, N.L., Muller, A.M. and Dzierzak, E.A. (1993) Nature 394, 64-67.
26. Owen, J.J., McLoughlin, D.E., Suniara, R.K. and Jenkinson, E.J. (2000) Curr. Top. Microbiol. Immunol. 251, 133-137.
27. Suniara, R.K., Jenkinson, E.J. and Owen, J.J. (2000) J. Exp. Med. 191, 1051-1056.
28. Globerson, A. (1966) J. Exp. Med. 123, 25-32.
29. Globerson, A. and Auerbach, R. (1965) Wistar Institute Monogr 4, 3-19.
30. Boyd, R.L., Tucek, C.L., Godfrey, D.I., Izon, D.J., Wilson, T.J., Davidson, N.J., Bean, A.G.D., Ladyman, H.M., Ritter, M.A. and Hugo, P. (1993) Immunol. Today 14, 445-459.
31. Von Gaudecker, B., Kendall, M.D. and Ritter, M.A. (1997) Microsc. Res. Tech. 38, 237-249.

32. Ropke, C., Van Soest, P., Platenburg, P.P. and Van Ewijk, W. (1995) Dev. Immunol. 4, 149-156.
33. Takeoka, Y., Chen, S.Y., Yago, H., Boyd, R., Suehiro, S., Shultz, L.D., Ansari, A.A. and Gershwin, M.E. (1996) Int. Arch. Allergy Immunol. 111, 5-12.
34. Boyd, R. and Chidgey, A. (2000) Immunol. Today 21, 472-474.
35. Georgopoulos, K., Bigby, M., Wang, J.H., Molnar, A., Wu, P., Winandy, S. and Sharpe, A. (1994) Cell 79, 143-156.
36. Wu, L., Nichogiannopoulou, A., Shortman, K. and Georopoulos, K. (1997) Immunity 7, 483-482.
37. Deftos, M.L. and Bevan, M.J. (2000) Current Opinion Immunol 12, 166-172.
38. Radtke, F., Ferrero, I., Wilson, A., Lees, R., Aguet, M. and MacDonald, H.R. (2000) J. Exp. Med. 191, 1085-1094.
39. Osborne, B. and Miele, L. (1999) Immunity 11, 653-663.
40. Wilson, A., Ferrero, I. MascDonald, H.R. and Radtke. F. (2000) J. Immunol. 165, 5397-5400.
41. Walker, L., Lynch, M., Silverman, S., Fraser, J., Boulter, J., Weinmaster, G. and Gasson, J. (1999) Stem Cells 17, 162-171.
42. Globerson, A. (1995) Int. Arch. Allergy Immunol. 107, 491-497.
43. Franceschi, C., Monti, D., Samsoni, P. and Cossarizza, A. (1995) Immunol. Today 16, 12-16.
44. Steinmann, G.G. and Muller-Hermelink, H.K. (1984) Immunobiol. 166, 45-52.
45. Douek, D.C. and Koup, R.A. (2000) Vaccine 18, 1638-1641.
46. Jamieson, B.D., Douek, D.C., Killian, S., Hultin, L.E., Scripture-Adams, D.D., Giorgi, J.V., Marelli, D., Koup, R.A. and Zack, J.A. (1999) Immunity 10, 569-575.
47. Miller. R.A. (1996) Science 273, 70-74.
48. Tamir, A., Eisenbraun, M.D., Garcia, G.G. and Miller, R.A. (2000) J. Immunol. 165, 1243-1251.
49. Utsuyama, M., Wakikawa, A., Tamura, T., Nariuchi, H. and Hirokawa, K. (1997) Mech. Ageing Dev. 93, 131-144.
50. Fulop, T., Jr., Gagne, D., Goulet, A.C., Desgeorges, S., Lacombe, G., Arcand, M. and Dupuis, G. (1999) Exp. Gerontol. 34, 197-216.
51. Effros, R.B. and Pawelec, G (1997) Immunol. Today 18, 450-454.
52. Effros, R.B. (2000) Vaccine 18, 1661-1665.
53. Mountz, J.D., Wu, J. and Hsu, H.C. (1997) Immunol. Rev. 160, 19-30
54. Thoman, M.L. Mech. (1995) Ageing Dev. 82, 155-170.
55. Timm, J.A. and Thoman, M.L. (1999) J. Immunol. 162, 711-717.
56. Kurashima, C., Utsuyama, M., Kasai, M., Ishijima, S.A., Konno, A. and Hirokawa, K. (1995) Int. Immunol. 7, 97-104.
57. Zlotnik, A. and Moore, T.A. (1995) Curr. Opin. Immunol. 7, 206-213.
58. Akashi, K., Kondo, M., Weissman, I.L. (1998) Immunol. Rev. 165, 13-28.
59. Soares, M.V.D., Borthwick, N.J., Maini, M.K., Janossy, G., Salmon, M. and Akbar, N.A. (1998) J. Immunol. 161, 5909-5917.
60. Haeryfar, S.M. and Berczi, I. (2001) Cell. Mol. Biol. (Noisy-le-grand) 47, 145-156.
61. Basch, R.S. (1990) Aging Immunol. Infect. Dis. 2, 229-235.
62. Hirokawa, K., Utsuyama, M., Katsura, Y. and Sado, T. (1988) Arch. Pathol. Lab. Med. 112, 12-21.
63. Fridkis-Hareli, M., Sharp, A., Abel, L. and Globerson, A. (1991) Thymus 18, 225-235.
64. Globerson, A., Eren, R., Abel, L. and Ben-Menahem, D. (1989) in: A.L. Goldstein (Ed.) Biomedical Advances in Aging, Plenum Press, Chapter 34, pp. 363-373.
65. Wang, Y., Tokuda, N., Tamechika, M., Hashimoto, N., Yamaguchi, M., Kawamura, H., Irifune, T., Choi, M., Awaya, A., Sawada, T. and Fukumoto, T. (1999) Histol. Histopathol. 14, 791-796.

66. Harrison, D.E. (1983) J. Exp. Med. 157, 1496-1504.

67. Eren, R., Zharhary, D., Abel, L. and Globerson, A. (1988) Cell. Immunol. 112, 449-455.

68. Harrison, D.E. (1980) Blood 55, 77-81.

69. Harrison, D.E., Jordan, C.T., Zhong, R.K. and Astle, C.M. (1993) Exp. Hematol. 21, 206-219.

70. Sharp, A., Kukulansky, T. and Globerson, A. (1990) Eur. J. Immunol. 20, 2541-2546.

71. Fridkis-Hareli, M., Abel, L. and Globerson, A. (1992) Immunology 77, 185-188.

72. Li, Y., Hisha, H., Inaba, M., Lian, Z., Yu, C., Kawamura, M., Yamamoto, Y., Nishio, N., Toki, J., Fan, H. and Ikehara, S. (2000) Exp. Haematol. 28, 950-960.

73. Small, M. and Weissman, I.L. (1996) Scand. J. Immunol. 44, 115-121.

74. Moriguchi, S. (1998) Biofactors 7, 77-86.

75. Mehr, R., Ubezio, P., Kukulansky, T. and Globerson, A. (1996) Aging: Immunol. Infec. Dis. 6, 133-140.

76. Ohkusu-Tsukada, K. Tsukada, T. and Isobe, K. (1999) J. Immunol. 163, 1966-1972.

77. Fink, P.J., Bevan, M.J. and Weissman, I.L. (1984) J. Exp. Med. 159, 436-451.

78. Michie, S.A. and Rouse, R.V. (1989) Thymus 13, 141-148.

79. Agus, D.B., Surh, C.D. and Sprent, J. (1991) J. Exp. Med. 173, 1039-1046.

80. Haynes, B.F. and Hale, L.P. (1998) Immunol. Res. 18, 61-78.

81. Mehr, R. and Perelson, A.S. (1997) Acquir. Immune. Defic. Syndr. Hum. Retrovirol. 14, 387-398.

82. Fridkis-Hareli, M., Mehr, R., Abel, L. and Globerson, A. (1994) Mech. Ageing Dev. 73:169-178.

83. Globerson, A., Kollet, O., Abel, L., Fajerman, I., Ballin, A., Nagler, A., Slavin, S., Ben-Hur, H., Hagay, Z., Sharp, A. and Lapidot, T. (1999) Exptl. Hematol. 27, 282-292.

84. Hollander, G.A., Wang, B., Nichogiannopoulou, A., Platenburg, P.P., Van Ewijk, W., Burakoff, S.J., Gutierrez-Ramos, J.C. and Terhorst, C. (1995) Nature 373, 350-353.

85. Bernstein, A. and Globerson, A. (1974) Cell. Immunol. 14, 171-181.

86. Scollay, R. and Godfrey, D.I. (1995) Immunol. Today 16, 268-273.

87. Kawamoto, H., Ohmura, K. and Katsura, Y. (1998) J. Immunol. 161, 3799-3802.

88. Solana, R. and Mariani, E. (2000) Vaccine 18, 1613-1620.

89. Globerson, A. (2001) in: G. Bock and J.A. Goode (Eds.) Aging vulnerability: causes and interventions. Novartis Foundation Symposium 235, Wiley, Chichester, pp85-100.

90. Peled, A. and Haran-Ghera, N. (1991) Virology 181, 528-535.

91. Andreu-Sanchez, J.L., Faro, J., Alonso, J.M., Paige, C.J., Martinez, C. and Marcos, M.A. (1990) Eur. J. Immunol. 20, 1767-1773.

92. Akashi, K., Richie, L.I., Miyamoto, T., Carr, W.H. and Weissman, I.L. (2000) J. Immunol. 164, 5221-5226.

93. Savino, W. and Dardene, M. (2000) Endocr. Rev. 21, 412-443.

94. Olsen, N.J. and Kovacs, W.J. (1996) Endocr. Rev. 17, 369-384.

95. Bodey, B, Bodey, B.Jr., Siegel, S.E. and Kaiser, H.E. (2000) Int. J. Immunopharmacol.22, 261-273.

96. Williams, N.S., Moore, T.A., Schatzle, J.D., Sivakumar, P.V., Zlotnik, A., Bennett, M. and Kumar, V. (1997) J. Exp. Med. 186, 1609-1614.

97. Kondo, M. and Weissman, I.L. (2000) Curr. Top. Microbiol. Immunol. 251, 59-65.

98. Moore, T.A. and Zlotnik, A. (1997) J. Immunol. 158, 4187-4192.

99. Updyke, L.W., Cocke, K.S. and Wierda, D. (1993) Mech. Ageing Dev. 69, 109-117.

100. Weekx, S.F.A., Snoeck, H.W., Offner, F., DeSmedt, M., Van Bockstaele, D.R., Nijs, G., Lenjou, M., Moulijn, A., Rodrigus, I., Berneman, Z.N. and Plum, J. (2000) Blood 95, 2806-2812.

101. Kawakami, N., Nishizawa, F., Sakane, N., Iwao, M., Tsujikawa, K., Ikawa, M., Okabe, M. and Yamamoto, H. (1999) J. Immunol. 163, 3211-3216.

102. Campbell, J.J., Pan, J. and Butcher, E.C. (1999) J. Immunol. 163, 2353-2357.
103. Ruiz, P., Wiles, M.V. and Imhof, B.A. (1995) Eur. J. Immunol. 25, 2034-2041.
104. Bodey, B., Bodey, B. Jr., Siegel, S.E. and Kaiser, H.E. (1997) *In Vivo* 11, 421-440.
105. Utsuyama, M. and Hirokawa, K. (1989) Mech. Ageing Dev. 47, 175-185.
106. Fitzpatrick, F.T., Kendall, M.D., Wheeler, M.J., Adcock, I.M. and Greenstein, B.D. (1985) J. Endocrinol. 106, R17-19.
107. Knyszynski, A., Adler-Kunin, S. and Globerson, A. (1992) Brain Behav. Immun. 6, 327-340.
108. Pearce, P., Khalid, B.A. and Funder, J.W. (1981) Endocrinol. 109, 1073-1077.
109. Seiki, K. and Sakabe, K. (1997) Arch. Histol. Cytol. 60, 29-38.
110. Sfikakis, P.P., Kostomitsopoulos, N., Kittas, C., Stathopoulos, J., Karayannacos, P., Dellia-Sfikakis, A. and Mitropoulos, D. (1998) Int. J. Immunopharmacol. 20, 305-312.
111. Olsen, J.N. and Kovacs, W.J. (1996) Endocr. Rev. 17, 369-384.
112. Cheleuitte, D., Mizuno, S. and Glowacki, J. (1998) Clin. Endocrinol. Metab. 83, 2043-2051.
113. Johnson, R.W., Arkins, S., Dantzar, R. and Kelley, K.W. (1997) Comp. Biochem. Physiol. A. Physiol. 116, 183-201.
114. Montecino-Rodriguez, E. Clark, R. and Dorshkind, K. (1998) Endocrinology 139, 4120-4126.
115. Marchetti, P., Scambia, G., Reiss, N., Kaye, A.M., Cocchia, D. and Iacobelli (1984) J. Ster. Biochem. 20, 835-839.
116. Staples, J.E., Gasiewicz, T.A., Fiore, N.C., Lubahn, D.B., Korach, K.S. and Silverstone, A.E. (1999) J. Immunol. 163, 4168-4174.
117. Amir-Zaltsman, Y., Mor, G., Globerson, A., Thole, H. and Kohn, F. (1993) Endoc. J. 1, 211-217.
118. Kohen, F., Abel, L., Sharp, A., Amir-Zaltsman, Y., Somjen, D., Luria, S., Mor, G., Thole, H. and Globerson, A. (1998) Dev. Immunol. 5:277-285.
119. Ashwell, J.D., Vacchio, M.S. and Galon, J. (2000) Immunol. Today 21, 644-646.
120. Head, G.M., Mentlein, R., VonPatay, B., Downing, J.E. and Kendall, M.D. (1998) Dev. Immunol. 6, 95-104.
121. Rinner, I., Felsner, P., Liebmann, P.M., Hofer, D., Wolfler, A., Globerson, A. and Schauenstein, K. (1998) Dev. Immunol. 6, 245-252.
122. Rinner, I., Globerson, A., Kawashima, K., Korstako, W. and Schauenstein, K. (1999) Neuroimmuno-modulation 6, 51-55.
123. Rinner, I., Kukulansky, T., Felsner, P., Skreiner, E., Globerson, A., Kasai, M., Hirokawa, K., Korsatko, W. and Schauenstein, K. (1994) Biochem. Biophys. Res. Commun. 203, 1057-1062.
124. Bulloch, K., McEwen, B.S., Nordberg, J., Diwa, A. and Baird, S. (1998) Ann. NY Acad. Sci. 840, 551-562.
125. Madden, K.S. and Felten, D.L. (2001) Cell. Mol. Biol. (Noisy-le-grand) 47, 189-196.
126. Egusa, Y., Fujiwara, Y., Syahruddin, E., Isobe, T. and Yamakido, M. (1998) Oncol. Rep. 5, 397-400.
127. Malaguarnea, M., Di Fazio, I., Vinci, E., Bentivegna, P., Mangione, G. and Romano, M. (1999) Panminerva Med. 41, 227-231.
128. Siegel, D.S., Desikan, K.R., Mehta, J., Singhal, S., Fassa, A., Munshi, N., Anaissie, E., Naucke, S., Ayers, D., Spoon, D., Vesole, D., Tricot, G. and Barlogie, B. (1999) Blood 93, 51-54.
129. Segal, R., Dayan, M., Globerson, A., Habot, B., Shearer, G. and Mozes, E. (1997) Mech. Ageing Dev. 96, 47-58.
130. Dayan, M., Segal, R., Globerson, A., Habut, B., Shearer, G.M. and Mozes, E. (2000) Exp. Gerontol. 35, 225-236.
131. Buchanan, J.P., Peters, C.A., Rasmussen, C.J. and Rothstein, G. (1996) Exp. Gerontol. 31, 135-144.

132. Rao, L.V., Cleveland, R.P., Kimmel, R.J., and Ataya, K.M. (1995) Am. J. Reprod. Immunol. 34, 257-266.

133. Chen, J., Astle, C.M. and Harrison, D. (2000) Exp. Hematol. 28, 442-450.

134. Miller, S. and Kearney, S.L. (1998) Lab. Animal Sci. 48, 74-79.

135. Brummendorf, T.H., Dragowska, W., Zijlmans, J.M.J.M., Thornbury, G. and Lansdorp, P.M. (1998) J. Exp. Med. 188, 1117-1124.

136. Traycoff, C.M., Orazi, A., Ladd, A.C., Rice, S., McMahel, J. and Srour, E.F. (1998) Exp. Hematol. 26, 53-62.

137. Vogel, G. (1999) Science 286, 238-239.

138. Gozes, Y., Umiel, T. and Trainin, N. (1982) Mech. Ageing Dev. 18, 251-259.

139. Globerson, A. (1986) in:. A. Facchini, J.J. Haaijman and G. Labo' (Eds.) Topics in Ageing Research in Europe, Vol. 9, pp. 9-15.

140. Harrison, D.E., Astle, C.M. and Stone, M. (1989) J. Immunol. 142, 3833-3840.

141. Morrison, S.J., Wandycz, A.M., Akashi, K., Globerson, A. and Weissman, I.L. (1996) Nature Medicine 2, 1011-1016.

142. Sharp, A., Zipori, D., Toledo, J., Tal, S., Resnitzky, P. and Globerson, A. (1989) Mech. Ageing Dev. 48, 91-99.

143. Sharp, A., Brill, S., Kukulansky, T. and Globerson, A. (1991) Ann. N.Y. Acad. Sci. 621, 251-269.

144. Morrison, S.J., Wright, D.E. and Weissman, I.L. (1997) Proc. Natl. Acad. Sci. USA 94, 1908-1913.

145. Rabinowich, H., Umiel, T. and Globerson, A. (1983) Transplantation 35, 40-48.

146. Peled, A., Petit, I., Kollet, O., Magid, M., Pnomaryov, T., Byk, T., Nagler, A., Ben-Hur, H., Many, A., Shultz, L., Lider, O., Alon, R., Zipori, D. and Lapidot, T. (1999) Science 283, 845-848.

147. Peled, A., Kollet, O., Ponomoyov, T., Petit, I., Franitza, S., Grabovsky, V., Slav, M.M., Nagler, A., Lider, O., Alon, R., Zipori, D. and Lapidot, T. (2000) Blood 95, 3289-3296.

148. Yu, S., Abel, L. and Globerson, A. (1997) Mech. Ageing Dev. 94, 103-111.

149. Mehr, R., Segel, L., Sharp, A. and Globerson, A. (1994) J. Theor. Biol. 170, 247-257.

150. Mehr, R., Fridkis-Hareli, M., Abel, L., Segel, L. and Globerson, A. (1995) J. Theor. Biol. 177, 181-192.

151. Norment, A.M. and Bevan, M.J. (2000) Semin. Immunol. 12, 445-455.

152. Bleul, C.C. and Boehm, T. (2000) Eur. J. Immunol. 30, 3371-3379.

153. Ogden, D.A. and Micklem, H.S. (1976) Transplantation 22, 237-293.

154. Ross, E.A.M., Anderson, N. and Micklem, H.S. (1982) J. Exp. Med. 155, 432-444.

155. Harrison, D.E., Astle, C.M. and Delaittre, J.A. (1978) J. Exp. Med. 147, 1526-1531.

156. Resnitzky, P., Segal, M., Barak, Y. and Dassa, C. (1987) Gerontology 33, 109-114.

157. Piacibello, W., Sanavio, F., Severino, A., Dane, A., Gammaitoni, L., Fagioli, F., Perissinotto, E., Cavalloni, G., Kollet, O., Lapidot, T. and Aglietta, M. (1999) Blood 93, 3736-3749.

158. Vaziri, H., Dragowska, W., Allsopp, RC., Thomas, T.E., Harley, C.B. and Lansdorp, P.M. (1994) Proc. Natl. Acad. Sci. USA 91, 9857-9860.

159. Chiu, C.P., Draowska, W., Kim, N.W., Vaziri, H., Yui, J., Thomas, T.E., Harley, C.B. and Lansdorp, P.M. (1996) Stem Cells 14, 239-248.

160. Engelhardt, M., Kumar, R., Albanell, J., Pettengell, R., Han, W. and Moore, M.A. (1997) Blood 90, 182-193.

161. Notaro, R., Cimmino, A., Tabarini, D., Rotoli, B. and Luzzatto, L. (1997) Proc. Natl. Acad. Sci. USA 94, 13782-13785.

162. Allsopp, R.C., Cheshier, S. and Weissman, I.L. (2001) J. Exp. Med. 193, 917-924.

163. Francz, P.I., Fridkis-Hareli, M., Abel, L., Bayreuther, K. and Globerson, A. (1992) Mech. Ageing

Dev. 64, 99-109.

164. Offner, F. and Plum, J. (1998) Leuk. Lymphoma 30, 87-99.

165. Offner, F., Kerre, T., De Smedt, M. and Plum, J. (1999) Br. J. Haematol. 104, 801-808.

166. Garcia-Ojeda, M.E., Dejbakhsh-Jones, S., Weissman, I.L. and Strober, S. (1998) J. Exp. Med. 187, 1813-1823.

167. Antica, M. and Scollay, R. (1999) J. Immunol. 163, 206-211.

168. Barda-Saad, M., Rozenszajn, L.A., Globerson, A., Zhang, A.S. and Zipori, D. (1996) Exp. Hematol. 24, 386-391.

169. Auerbach, R., Wang, S.J., Yu, D., Gilligan, B. and Lu, L.S. (1998) Dev. Comp. Immunol. 22, 333-338.

170. Hirayama, F., Aiba, Y., Ikebuchi, K., Sekiguchi, S. and Ogawa, M. (1999) Blood 93, 4187-4195.

171. Pawelec, G., Muller, R., Rehbein, A., Hahnel, K. and Ziegler, B.L. (1998) J. Leukoc. Biol. 64, 733-739.

172. Sharp, A., Kukulansky, T., Malkinson, Y. and Globerson, A. (1990) Mech. Ageing Dev. 52, 219-233.

173. Ben-Yehuda, A., Friedman, G., Wirtheim, E., Abel, L. and Globerson, A. (1998) Mech. Ageing Dev. 102, 239-247.

174. Aspinall, R. and Andrew, D. (2000) Vaccine 18, 1629-1637.

175. Wu, Q., Tidmarsh, G.F., Welch, P.A., Pierce, J.H. Weissman, I.L. and Cooper, M.D. (1989) J. Immunol. 143, 3303-3308.

176. Sherwood, P.J. and Weissman, I.L. (1990) Int. Immunol. 2, 399-406.

177. Miller, R.A. and Harrison, D.E. (1985) J. Immunol. 134, 1426-1429.

178. Weindruch, R.H. and Suffin, S.C. (1980) J. Gerontol. 35, 525-531.

179. Miskin, R., Masos, T., Yahav, S., Shinder, D. and Globerson, A. (1999) Neurobiology of Aging 20, 555-564.

180. George, A.J.T. and Ritter, M.A. (1996) Immunol. Today 17, 267-272.

CHAPTER 4

PLASTICITY OF ADULT BONE MARROW STEM CELLS

KAREN J. CHANDROSS and ÉVA MEZEY

Table of contents

1.   Introduction

In mammals, including humans, newly differentiated cells are continuously generated from stem cells throughout development. In the adult, stem cells are found in different organ systems where they can contribute to the replacement of cells lost to physiological turnover, injury, or disease. Until recently the idea that different stem cell populations exist, each unique to the organ or tissue that they occupy, remained unchallenged. There was little evidence that stem cells in one organ were capable of differentiating into cells characteristic of another. Bone marrow stem cells were thought to be restricted in their potential to become mature stromal or hematopoietic cells. However, an increasing

Stem Cells: A Cellular Fountain of Youth.   Ed. by Mark P. Mattson and Gary Van Zant. 73 — 95

number of studies demonstrate that stem cells in adult organs exhibit greater plasticity than previously recognized. When taken from their residence bone marrow-derived stem cells develop characteristics that typify brain, muscle, liver, and endothelial cells. Analogously, brain-derived stem cells exhibit characteristics of hematopoietic and muscle cells. The fate of a cell is therefore likely to be dictated in part by the local environment. This chapter reviews the evidence for bone marrow stem cell plasticity, focusing on the differentiation of these cells into different neural cell types and their potential therapeutic value in the treatment of central nervous system (CNS) injury and disease.

2.   Adult stem cells

2.1   General biology

The term "stem cell" describes a cell that can self-renew (produce identical stem cells) and, through asymmetric division, also gives rise to more differentiated daughter cells, known as progenitor cells. As a progenitor cell differentiates, the number of subsequent cell divisions and its lineage potential becomes increasingly restricted. In adult mammals, somatic stem cells have been identified in a number of tissues, such as blood, bone marrow, central and peripheral nervous system, retina, cornea, skin, muscle, liver, and intestinal crypts (Table 1). Until recently, it was generally thought that mammalian somatic stem cells were multipotent, meaning that their differentiation potential was restricted to the resident cells in the tissue they occupied. Several studies now suggest that stem cells have a broad spectrum of differentiation potential and may be capable of changing their phenotypic and functional properties depending on their location (Figure 1, Table 2).

Table 1.   Possible sources of adult stem cells and clinical approaches for the treatment of nervous system injury or disease

Sources of adult mammalian stem cells:

Blood/bone marrow	[see text]
Neural Tissue (CNS and PNS)	[76, 92, 97, 98, 99, 100]
Retina	[125]
Cornea	[126]
Skin	[127]
Muscle	[30]
Liver	[48]
Intestinal crypts	[128, 129]

Avenues of clinical treatment:

•Transplant stem cells after they have been pushed down specific neural lineages
•Transplant genetically altered stem cells pre-programmed to progress down a restricted lineage
•Identify factors that regulate the expansion and differentiation of endogenous or transplanted stem cells

Table 2. Published studies demonstrating bone marrow stem cell plasticity. Adult stem cells retain a certain degree of plasticity in that they differentiate into mature cells found in the organ where they reside. In a broader sense, adult stem cells can give rise to progeny derived from the same germ layer. For example, bone marrow-derived progenitor cells (mesoderm) differentiate into myogenic cells and muscle precursor cells (mesoderm) that can reconstitute the hematopoietic lineage. Stem cells may exhibit an even greater degree of plasticity in that those present in distinct organs can give rise to cells characteristic of different germ layers. There are an increasing number of studies using different animal models that suggest that bone marrow-derived stem cells become liver oval cells (endoderm) and neural cells (neuroectoderm).

Bone marrow transition	Model system	References
Brain		
Bone marrow → astrocytes/microglia	Intravenous transplantation of acutely isolated and cultured stem cells into irradiated mice.	[58]
Stromal cells → astrocytes	Transplantation of cultured adult marrow stromal cells into corpus striatum of mice. Transplantation of cultured marrow stromal cells into lateral ventricle of mice.	[63] [64]
Bone marrow cells → neurons	Intraperitoneal transplantation of acutely isolated bone marrow cells into PU.1 knockout mice. Intravenous transplantation of acutely isolated bone marrow cells into irradiated mice.	[65] [79]
Stromal cells → neural cells	Neurons/astrocytes: *In vitro*, mouse and human cells BME, retinoic acid, BDNF). Neurons: *In vitro*, rat and human cells, rat lines (DMSO, BME).	[86] [87]
Muscle		
Bone marrow cells → muscle	Intramuscular transplantation into immunodeficient, *scid/bg*, mice.	[25, 26]
Bone marrow stromal cells → myocytes/muscle	Xenotransplantation of human stromal cells into fetal sheep	[28]
Bone marrow hematopoietic stem cells → muscle	Intravenous transplantation into *mdx* mice.	[27]
Endothelial cells		
Peripheral blood hematopoietic stem cells → endothelial cells	*In vitro*, human (bovine serum and brain extract).	[34]
Various sources of hematopoietic stem cells → endothelial cells	*In vivo*, xenograft of human into irradiated canines. *In vitro* (bFGF, IGF-1, VEGF).	[40]
Liver		
Bone marrow cells → hepatic oval cells/hepatocytes	Transplantation.	[47, 49, 130]
Bone marrow hematopoietic stem cells → hepatocytes	Transplantation into lethally irradiated mice.	[50]

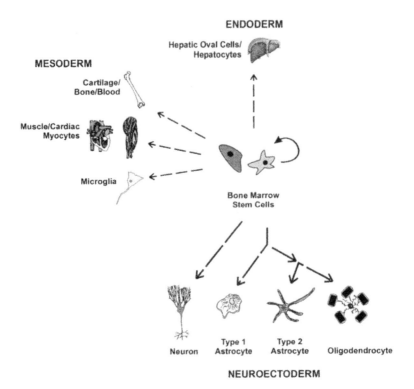

Figure 1.   Adult bone marrow-derived stem cells can differentiate into a variety of organ-specific cell types, including cells found in the brain, muscle, lung, and liver (see Table 2).

Mechanistic models for fate changes were proposed for the transdetermination of Drosophila imaginal discs, the primordia of the adult exoskeleton and appendages [1]. After regenerative cell divisions, imaginal disc cells changed from one state of determination to another, initiating a pathway of differentiation leading to structures other than those corresponding to the initial state of determination. For example, an antennal imaginal disc differentiated into a leg imaginal disc. This process has been described as transdifferentiation, which refers to the ability of cells committed to a particular fate to acquire a new identity by abandoning one set of lineage-specific genes and adopting the expression pattern of another differentiated cell type. The decision for a stem cell to change its course of differentiation, however, is subject to its location in space and time. In this sense, switches in fate decisions do not necessarily represent transdifferentiation and, instead, may reflect a wide range of developmental potentials expressed by adult stem cells under certain conditions in the absence of otherwise inhibitory cues.

There is evidence that stem cells change their properties over time. In the immune system, fetal hematopoietic stem cells have different antigenic properties and exhibit broader lineage potentials than do adult hematopoietic stem cells [2]. For example, adult hematopoietic stem cells were not capable of making certain classes of T cells even when transplanted into the fetal environment [3] and were less efficient at producing B cells and erythrocytes than fetal cells [4, 5]. In humans, fetal bone marrow had a higher number of stem cells, lower immunological reactivity, and higher clonogenic/proliferative potential than did adult bone marrow [6]. Changes in the properties of stem cells are not unique to the immune system. Qian and colleagues used an *in vitro* model to follow the differentiation of individual mouse stem cells derived from the cortex during the period of neurogenesis [7]. In this study, single progenitor cells from embryonic day 10 generated neuroblasts before glioblasts, consistent with previous work demonstrating that *in vivo* neurogenesis proceeds gliogenesis. The authors concluded that the program that initiates neuronal differentiation prior to glial cell differentiation is intrinsically encoded within CNS stem cells.

Several intrinsic mechanisms may lead to heterogeneity among stem cells. Intrinsic changes in telomere length [8, 9, 10, 11], the level of cyclin-dependent kinase inhibitor P27 [12], or expression of growth factor receptors [13, 14] may influence stem cell specification decisions. Local paracrine and autocrine factors and feedback mechanisms involving different cell types may modulate stem cell fate decisions beginning very early in development, thereby leading to heterogeneity among stem cells. Thus, stem cells from different organ systems may exhibit subtle differences in gene expression and developmental potential, underscoring our need to rigorously define cellular classifications.

An increasing number of studies suggest that even if their properties change, stem cells remain adaptable in nature throughout ontogeny. For example, under the appropriate conditions, adult and fetal stem cells exhibited similar growth characteristics and regulatory genes, and differentiated into the same neural cell types [15]. It has been suggested that the differentiated state of a cell is reversible and requires continuous regulation to maintain the balance of factors present in a cell at any given time [16]. The cloning of Dolly [17] underscored the potential for previously silent mammalian genes to become activated in adult nuclei. Moreover, there are now a number of papers demonstrating that under physiologic conditions, stem cells from one organ are capable of differentiating into cells characteristic of another organ (Figure 1, Table 2). Recent studies obscure distinctions between tissues derived from the three embryonic germ layers (compare Figure 1 and Table 2 to Table 3) and provide the foundations for investigating new approaches to treating human disease. Thus, although the intrinsic properties of stem cells may undergo subtle changes over time, developmental plasticity is not restricted to the embryonic environment.

Table 3. Embryonic progenitors of differentiated tissues.

Embryonic tissue	Differentiated tissue
Endoderm	Thymus; thyroid, parathyroid glands; larynx, trachea, lung; urinary bladder, vagina, urethra; gastrointestinal organs (liver, pancreas); lining of the GI tract; lining of the respiratory tract
Mesoderm	Bone marrow; adrenal cortex; lymphatic tissue; skeletal, smooth, and cardiac muscle; connective tissues (including bone, cartilage); urogenital system; heart and blood vessels (vascular system)
Ectoderm	Skin; neural tissue (neuroectoderm); adrenal medulla; pituitary gland; connective tissue of the head and face; eyes; ears

3.   The plastic nature of adult bone marrow stem cells

3.1.   Bone marrow stem cells

Circulating blood cells have a relatively short lifespan and are continually replaced. Hematopoietic stem cells in the bone marrow provide a constant source of progenitor cells for red blood cells, platelets, monocytes, granulocytes, and lymphocytes throughout the life of an individual and can reconstitute the entire hematopoietic system after bone marrow transplantation [18]. Through the work of many researchers over the past several decades, cell surface and intracellular properties of human and murine hematopoietic stem cells have been established [reviewed in 19].

The first evidence that bone marrow contains precursor cells for non-hematopoietic mesenchymal tissues came from the work of Friedenstein and co-workers [20]. The precursor cells were initially referred to as colony-forming-units-fibroblasts and later as mesenchymal stem cells, or marrow stromal cells. Stromal stem cells are present throughout the lifespan of an organism and give rise to the support cells of the bone marrow, including colonies of osteoblasts, chondroblasts and adipocytes [20, 21, 22].

Several ideas have been put forth to explain why marrow contains cells that have the potential to differentiate into a variety of mesenchymal cells [reviewed by 23]. For example, marrow contains a complex array of thin spicules of trabecular bone, which is similar to other bone in that it continually undergoes remodeling. Therefore marrow may include osteoblast precursors that were eluted from trabecular bone or the inner surface of the bone itself. Chondrocyte differentiation might be associated with the process of fracture repair, although this process continues to occur in bone that lacks marrow. The potential of stromal stem cells to differentiate into adipocytes may be related to the finding that marrow is in part replaced with adipose tissue with aging.

Recent studies further challenged the notion of unilateral development in the immune system by demonstrating that hematopoietic and stromal cells can undergo transitions within and between individual embryonic germ layer classifications. For example, mesoderm-derived bone marrow cells can give rise to muscle and endothelial cells (derived from mesoderm), but can also become cells characteristic of the liver (derived from endoderm), and brain (derived from neuroectoderm) (Figure 1, Tables 2 and 3).

3.2.   Bone marrow to muscle

A decade ago, it was observed that the cytoskeleton of cultured bone marrow stromal cells contained the alpha smooth muscle isoform of actin [24]. The intermediate filaments contained vimentin but no desmin filaments, properties that are shared with vascular smooth muscle cells. Subsequent research efforts showed that cultured bone marrow cells converted to a myogenic pathway when transplanted into chemically induced degenerating muscle [25, 26]. Ferrari and colleagues [25] transplanted genetically tagged bone marrow cells into the damaged muscles of the *scid/bg* immunodeficient mice, which lack functional B and T cells and have intrinsically low natural killer cell activity. The transplanted cells migrated into the areas of induced muscle degeneration, underwent myogenic differentiation, and participated in the regeneration of the damaged fibers by giving rise to differentiated muscle fibers. These results suggested that cells were recruited by long-range signals that originated from degenerating tissue and subsequently accessed the damaged muscle from the peripheral circulation. Following transplantation bone marrow-derived cells took longer to differentiate into muscle than committed myogenic precursors. These results suggested that in this environment bone marrow-derived progenitors underwent distinct differentiation processes than the endogenous cell population.

Both hematopoietic and stromal bone marrow stem cell populations are capable of differentiating into muscle [25, 27, 28]. In response to 5-azacytidine and amphotericin B, cultured stromal stem cells differentiated into myoblasts that fused into rhythmically beating myotubes [29]. Transplantation studies showed the localization of hematopoietic stems to damaged muscle in mice [27]. In this study, hematopoietic stem cells carrying a wild-type dystrophin gene were transplanted intravenously into *mdx* mice, an animal model of Duchenne's muscular dystrophy. Following lethal irradiation, mice received tail vein injections of purified hematopoietic bone marrow cells from wild type mice. At 12 weeks after bone marrow transplantation, about 4% of the muscle fibers of an individual mouse expressed dystrophin, with 10-30% of these cells being derived from the donor, indicating that bone marrow cells can integrate into and restore function to damaged muscle. Further studies are needed to identify ways in which to increase the level of engraftment so that it would be clinically useful in the treatment of muscular diseases.

Interestingly, adult muscle contains a population of stem-like cells that exhibit characteristics of bone marrow-derived hematopoietic stem cells [30]. Six to twelve weeks after transplantation into irradiated mice, cultured skeletal muscle stem cells showed high-level engraftment representing all major adult blood lineages. The authors concluded that these putative stem cells may be identical to muscle satellite cells, some of which lack myogenic regulators and could be expected to respond to hematopoietic signals.

3.3.   Bone marrow to endothelial cell

Vascularization is the differentiation of mesodermal precursor cells to angioblasts that become endothelial cells, which in turn give rise to the primitive capillary network [31]. Angiogenesis, vascularization that occurs postnatally, is thought to arise from the proliferation, migration, and remodeling of circulating endothelial cells derived from blood vessels [32, 33].

Human mononuclear progenitor cells were purified from peripheral blood by the expression of CD34 and Flk-1 [34], antigens that are expressed in both angioblasts and hematopoietic stem cells [35, 36, 37, 38, 39]. When these cells were grown in the presence of bovine serum and brain extract, they differentiated into mature endothelial cells *in vitro*. In a subsequent study [40], CD34 positive hematopoietic stem cells were isolated from human umbilical cord, fetal liver, bone marrow, and granulocyte colony stimulating factor-mobilized peripheral blood and then cultured in a combination of basic fibroblast growth factor (bFGF), vascular endothelial growth factor (VEGF), and insulin-like growth factor-1 (IGF-1). After 15 to 20 days *in vitro*, the cells became indistinguishable from mature endothelial cells. After grafting the CD34 positive human bone marrow cells into the descending thoracic aortas of irradiated canine recipients, donor cells differentiated into endothelial cells and formed extensive monolayers. Moreover, adult human bone marrow-derived endothelial precursors improved cardiac function in rats following experimental myocardial infarction [41]. Transplanted progenitor cells mobilized to the infarct and induced both vasculogenesis and angiogenesis, thereby protecting hypertrophied myoctyes against apoptosis and preventing remodeling (e.g., collagen deposition). Taken together, these studies indicate that bone marrow-derived stem cells may contribute to the formation of new blood vessels and help to prevent heart failure in humans.

3.4.   Bone marrow to liver

Oval cells give rise to two epithelial cell types present in the liver, the ductal cells and hepatocytes [42, 43, 44]. Oval cells have similar phenotypic traits to bone marrow stem cells [45, 46, 47] and hematopoietic stem cells, which arise from the bone marrow, are present in the liver [48]. Transplantation studies were utilized to trace the origin(s) of cells that repopulated the liver and test the possibility that hepatic oval cells may arise from a cell population originating in the bone marrow. Bone marrow cells that were transplanted into lethally irradiated mice constituted about 0.14% of the total hepatocytes and 0.1% of the total oval cells [49], providing evidence that cell(s) associated with the bone marrow may act, under certain conditions, as liver progenitor cells.

In a separate study, highly enriched hematopoietic stem cells were injected intravenously into fumarylacetoacetate hydrolase (FAH)-deficient mice, an animal model of human tyrosinemia type 1 liver disease [50]. Mutant mice have progressive liver failure and renal tubular damage unless treated with 2-(2-nitro-4-trifluoro-methylbenzyol)-1, 3-cyclohexanedione [51]. FAH-deficient mice were used as bone marrow recipients because transplanted wild-type hepatocytes vigorously repopulated the mutant liver [52]. As few as fifty transplanted adult hematopoietic stem cells were able to reconstitute hematopoiesis and gave rise to sufficient hepatocytes to restore liver function.

3.5.   Bone marrow to brain

3.5.1.   *Bone marrow to brain: in vivo evidence*

It is now widely accepted that circulating T-cells, B-cells, macrophages, and microglia enter the brain and migrate throughout the CNS [53, 54, 55, 56, 57, 58]. Genetic mouse models have been used to examine the migration of bone marrow cells into the CNS during disease. The twitcher mouse is a model of human globoid cell leukodystrophy, which is caused by the mutation of the galactosylceramidase gene. The toxic metabolite, psychosine, accumulates in both the central and peripheral nervous systems and results in the death of oligodendrocytes and subsequent demyelination [59, 60]. Rhodamine isothiocyanate was used to mark and detect infiltrating cells in the diseased nervous system [61]. Prior to one month after intraperitoneal injection of this fluorescent dye into twitcher mice, labeled cells were restricted to the leptomeninges, choroid plexus, and supra ependymal regions of the ventricles, and were not present in CNS parenchyma. However, one month after injection labeled cells were present throughout the parenchyma of adult twitcher mice, which contrasted with the absence of cells within the CNS of wild type mice. In another study, bone marrow cells obtained from enhanced green fluorescent protein (GFP)-expressing transgenic mice with normal galactosylceramidase activity were transplanted into mice [62]. Wild type and twitcher recipients exhibited engraftment of splenic monocytes/macrophages within a month, with slower infiltration into the lung, kidney, and liver. Again, prior to one month after transplantation donor cells were localized to the perivascular and leptomeningeal regions of wild type mice. In twitcher chimeric mice, significant numbers of donor cells infiltrated into the CNS after one month and were present throughout the white matter of the spinal cord, brain stem, and cerebrum. Additionally, cells that co-labeled with microglial antigenic markers were evenly distributed throughout white and gray matter. By contrast, labeled cells were not detected in the CNS parenchyma of normal mice.

Taken together, these studies suggest that there is a delay in the migration of bone marrow cells into the CNS parenchyma. Moreover, an increased number of hematogenous cells may infiltrate the nervous system after injury, possibly in response to injury-induced changes in chemokine and cytokine expression and/or alterations in the properties of the blood-brain, blood-cerebrospinal fluid, or cerebrospinal fluid-brain barriers, which separate the periphery from the CNS.

Eglitis and Mezey demonstrated that when bone marrow cells (a mixture of hemato-poietic and stromal cells) were injected into the tail vein of irradiated recipient mice, they entered the brain and became astrocytes, as well as microglia [58]. These were the first published studies demonstrating that mesoderm-derived bone marrow cells can give rise to astrocytes, which are thought to be of neuroectodermal origin. Subsequent transplantation studies reconfirmed these findings. When cultured mesenchymal stem cells from the bone marrow stroma were transplanted into the lateral ventricle or striatum of recipient mice, these cells migrated throughout the brain and differentiated into astrocytes, as determined by vimentin and Glial fibrillary acidic protein (GFAP) immunoreactivity [63, 64]. After engraftment, donor cells no longer expressed collagen and fibronectin, antigenic markers typical of marrow stromal cells in culture.

Our most recent work using mice homozygous for a mutation in the PU.1 gene as bone marrow recipients suggests that some of the neurons present throughout the adult nervous system may arise from cells present in adult bone marrow [65]. PU.1 is a member of the E26 avian erythroblastosis virus transformation-specific (ETS) DNA binding domain family of transcription factors, which is expressed exclusively in cells of the hematopoietic lineage. Mice homozygous for a disruption in the PU.1 binding domain lack macrophages, osteoclasts, neutrophils, mast cells, B and T cells at birth [66, 67]. PU.1 null animals are born alive but die of severe septicemia (due to the lack of macrophages) within 48 hr after birth. Autopsies show that the bones of these mice have an osteopetrotic phenotype caused by an insufficient number of osteoclasts. However, when PU.1 knockout mice receive bone marrow transplants at birth, they develop normally, are fertile, and live into adulthood. There are several advantages to using PU.1 mice as transplant recipients; 1) they are immunodeficient, thereby obviating the need to irradiate recipients prior to transplantation in order to avoid immune-related host responses and 2) since all marrow cells are derived from donors, the number of cells that migrate into the nervous system can be accurately assessed.

In our studies [65], male donor cells were identified in the nervous system of female recipients by non-radioactive fluorescent *in situ* hybridization of the Y chromosome combined with immunohistochemistry, to determine the phenotype of individual donor cells. GFAP was used to identify mature astrocytes [68] and subventricular zone stem cells [69] and NeuN, a nuclear transcription factor localized exclusively to neuronal nuclei [70, 71, 72], was used to examine the co-localization of the Y-chromosome to neurons. Prior to transplantation, isolated donor cells did not express nestin (a cytoskeletal marker found in neural stem cells, [73]), NeuN, NG2 (chondroitin sulfate proteoglycan marker present in oligodendrocytes, [74]) or O4 (oligodendrocyte-specific sulfitide and seminolipid, [75]). After growing in culture for several weeks in the presence of platelet-derived growth factor, bone marrow stem cells (hematopoietic and stromal) expressed nestin and after one month greater than 50% of the cells were nestin positive [65].

Marrow-derived donor cells (i.e., Y chromosome positive) were present in the CNS of all of the transplanted mice examined. The Y chromosome bearing cells were evenly distributed throughout the different brain regions, in both white and gray matter. Up to 5% of all cells were derived from the donor. Interestingly, a large number of Y chromosome positive nuclei were present in cells within the choroid plexus of the lateral ventricle, ependyma (a single cell layer lining the ventricular system), and subventricular zone (Figure 2A, B). Thus the choroid plexus may provide a conduit for circulating cells to enter the brain. By contrast, very few donor cells were found associated with capillaries within the parenchyma.

Recent studies showed that GFAP immunoreactive cells within the subventricular zone were neural stem cells that gave rise to neurons and glial cells [76]. In our studies, many Y positive donor cells within the subventricular zone were associated with GFAP immunoreactivity [77]. More sensitive assays (better markers and confocal z series analyses) are needed to determined whether CNS stem cells can arise from the periphery. As we previously published [58, 65], many Y-chromosome positive cells present throughout the brain were indistinguishable from endogenous astrocytes (Figure 2C). Similar results were obtained using two different mouse models (irradiated mice *versus* PU.1 knockout mice) and different entry routes for the donor cells (tail vein *versus* intraperitoneal) [58, 65], thereby reinforcing the idea that bone marrow-derived cells can in fact become astrocytes.

Even more striking, 1-2% of NeuN immunoreactive neurons were Y-positive (Figure 3), as confirmed by z-series confocal imaging [65] and we recently identified Y positive donor cells in CNS white matter that were morphologically indistinguishable from oligodendrocytes. To determine whether bone marrow cells that had integrated into the parenchyma expressed markers of mature oligodendrocytes, mice were analyzed one month after receiving intraventricular injections of bone marrow stem cells 3 days after birth (K.J. Chandross, E.M. Mezey, A. Guzik, K. Auble, and L.D. Hudson, unpublished data). Bone marrow cells were isolated from adult transgenic mice that ubiquitously

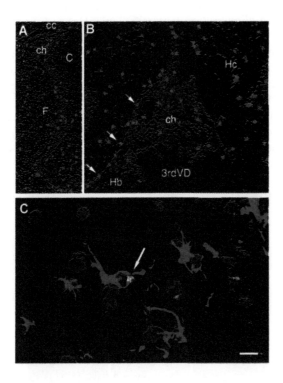

Figure 2.   Bone marrow cells can migrate into the CNS ventricular system, integrate into the subventricular zone and parenchyma, and become astrocytes. Donor cells were identified by the presence of the Y chromosome in PU.1 knockout female recipient mice two months after receiving intraperitoneal injections of acutely isolated male bone marrow cells at birth. PU.1 mice were used as recipients because all cells that incorporate into the central nervous must be derived from the donor cells, thereby optimizing the number of tagged cells that can be followed. **(A and B)** Coronal images of the lateral ventricle and dorsal aspect of the third ventricle, showing a high density of Y positive cells in the choroid plexus (ch) and ependyma and subependymal regions (arrows). The findings suggest that these regions may be portals of entry into the brain for bone marrow-derived cells. Methods: The Y chromosome riboprobe was labeled using ^{35}S-UTP and sections were emulsion coated, developed after one week, and then stained with 4, 6-diamidino-2-phenylindole dihydrochloride (DAPI, blue), to visualize nuclei. The autoradiographic grains were photographed in darkfield illumination through a red filter. **(C)** Y chromosome and GFAP double positive astrocyte in the cerebral cortex (arrow) of a female recipient mouse. Method: The Y chromosome was using a non-radioactively labeled riboprobe followed by an amplification using FITC-Tyramide (green). Astrocytes were identified using a CY-3 (red) conjugated anti-GFAP antibody. The cell nuclei were labeled with DAPI (blue). Abbreviations: C, caudate; cc, corpus callosum; ch, choroid plexus; F, fimbria hippocampi; Hb, medial habenula; Hc, hippocampus; 3rdVD, dorsal portion of the third ventricle. Scale bar represents (A) 100 μm; (B) 60μm, and (C) 10 μm.

expressed the enhanced green fluorescent protein (GFP, [78]). Within 3 weeks after transplant, we identified donor cells in the corpus callosum that were immunoreactive for both GFP and the oligodendrocyte-specific antigenic marker, myelin basic protein (Figure 4, top four panels). It will be necessary to look at later time points since only a few oligodendrocytes could be detected prior to 30 days after transplant.

We did not observe an increased density of Y chromosome positive cell nuclei in neurogenic regions, including the subventricular zone, olfactory migratory region, and hippocampus [65]. However, in contrast to neural stem cells, mesodermal stem cells also differentiate into microglia [57]. Therefore, since all microglia in the recipient animals are also Y chromosome positive, any regional differences in the number of Y positive neurons would be masked by the significantly higher number of homogeneously distributed microglia.

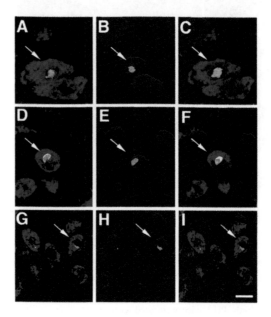

Figure 3.　Bone marrow cells can become CNS neurons. Colocalization of the Y chromosome to NeuN immunopositive neurons in the cortex of female PU.1 bone marrow recipients. **(A)** and **(D)** Fluorescent images of the olfactory cortex. The same fields were overlaid to show the colocalization of Y chromosome hybridization (detected using Tyramide-FITC conjugate, green) to NeuN immunopositive (red) nuclei. **(B)** and **(E)** represent the identical fields to A and D, respectively, taken through the UV filter to show all (not only neuronal) DAPI positive cell nuclei (blue) overlaid with the Y chromosome images. Note that each Y chromosome was associated with a cell nucleus. **(C)** and **(F)** Triple color images of the same fields. **(G-I)** Confocal images (one level out of a Z series) of a double labeled Y chromosome and NeuN positive neuron in the parietal cortex (1.2 mm behind the bregma) of a 1-month-old female PU.1 knockout mouse transplanted at birth with male bone marrow. The confocal microscope examined one slice of the section and determined that the Y chromosome was in the same plane and was associated exclusively with the neuronal nucleus shown in the image. Arrows identify donor cells that were NeuN immunopositive. Scale bar: A-F, 10 μm; G-I, 20μm.

In an independent study, Brazelton and colleagues [79] showed that bone marrow-derived cells administered by intravascular injection, entered the brain of lethally irradiated adult mice and expressed neuron-specific genes. Adult donor cells were easily detected in non-transgenic recipients because they were isolated from transgenic mice ubiquitously expressing the GFP protein [GFP reference 78]. Flow cytometry comparisons of the donor cells derived from the bone marrow and brain of transplanted animals revealed striking differences. Most of the tagged cells isolated from the recipient bone marrow expressed CD45, which identifies all nucleated mature blood lineages [80]. By contrast, up to 20% of the cells isolated from the brains of recipient mice lacked both CD45 and CD11b, a marker expressed by all myelo-monocytic cells, including microglial cells in the CNS [81, 82], suggesting that upon entering the brain bone marrow-derived cells assumed novel phenotypes. Laser scanning confocal microscopy analyses of the olfactory bulb confirmed that individual donor cells had integrated in the microenvironment and coexpressed GFP and the neuronal proteins [70], NeuN, 200-kD isoform of neurofilament, and class III β-tubulin (TuJ1). By 8 to 12 weeks after transplantation, approximately 0.2-0.3% of the total number of neurons were derived from the bone marrow. Many donor cells were immunopositive for both the 200 kD isoform of neurofilament (found in mature neurons) and class III β-tubulin, lending support to the idea that bone marrow cells differentiated into neurons. The majority of cells expressing neuronal antigens had short nonbranching extensions, reminiscent of immature neurons. This might be expected in the olfactory bulb since class III β-tubulin expressing neuronal progenitor cells migrate into this region from the subventricular zone throughout life [83, 84, 85]. However, it is unclear why these cells expressed markers of mature neurons, underscoring the need for further characterization.

3.5.2. *Bone marrow to brain: in vitro evidence*

In vitro models can provide a powerful approach for examining the properties of individual cells. Two independent tissue culture studies suggested that stromal progenitor cells found in bone marrow were induced to become neurons [86, 87]. In the first study [86], bone marrow stromal cells were enriched through negative selection (the elimination of hematopoietic stem cells). In the presence of retinoic acid and brain derived neurotrophic factor, 0.5% of the negatively selected cells expressed the neuronal antigen, NeuN. Co-culturing stromal cells on a layer of mouse fetal midbrain cultures for one week increased the number of NeuN positive cells at least two-fold, indicating that the cellular microenvironment may augment the differentiation of bone marrow stromal cells. However, *in vitro* NeuN expression is not restricted to neuroectoderm-derived cells and after several weeks in culture these cells never expressed markers of mature neurons. In a separate study, Woodbury and colleagues cultured stromal cells in the presence of β-mercaptoethanol or dimethylsulfoxide and relatively high concentration steroids and forskolin [87]. The authors concluded that under these culture conditions, stromal cells expressed neuronal antigens within thirty minutes, a finding that has not yet been observed or reconfirmed by others.

Due to the lack of well-characterized nuclear oligodendrocyte-specific antigens, the actual localization of the Y chromosome to oligodendrocytes *in situ* is difficult. For this reason we turned to a tissue culture system to determine whether Sca1 positive (hematopoietic) and/or Sca1 negative (stromal) stem cells could give rise to oligodendrocytes. After 24 hours in culture, bone marrow cells did not express the oligodendrocyte antigenic marker, NG2. However, within 30 days 2% of the progenitor cells initially isolated by sorting for the presence of Sca1 (Figure 4, bottom two panels) expressed NG2 and looked similar to pro-oligodendrocytes (K.J. Chandross, A. Guzik, E. Mezey, and L.D. Hudson, unpublished data). These preliminary studies suggest that bone marrow stem cells can give rise to oligodendrocytes. Our current efforts are focused on optimizing the conditions for pushing bone marrow stem cells down the oligodendrocyte lineage and transplanting genetically tagged progenitor cells into mouse models of demyelinating disease.

Although reports that bone marrow stem cells can become neurons and oligodendrocytes are highly provocative, they should be interpreted cautiously. Additional studies are needed to determine whether neural cells can be obtained *in vitro* in response to physiological signals and whether the resulting cells exhibit the appropriate functional characteristics.

3.5.3.   *Bone marrow to brain: an optimistically cautious note*

In several *in vitro* studies [65, 86, 87], bone marrow stem cells were induced to express nestin, a marker for neuroectoderm-derived stem cells. However, nestin is also expressed outside of the nervous system in myogenic cells, hepatic stellate cells, epithelial cells of the developing lens, and in newly formed endothelial cells of the extra- and intraembryonic blood vessels [88, 89], underscoring the need to identify a panel of markers that better identify specific cell types. It remains to be determined whether bone marrow cells transiently express markers in common with neural stem cells or whether these cells are in fact more closely related than previously recognized.

Where do new neurons come from in adult mammals? Taken together, the above studies suggest that the blood may supply the brain with a source of neural cells, including neural stem cells, astrocytes, neurons, and oligodendrocytes. Further studies are needed to determine whether in the adult stem cells can migrate between different organ systems. The use of tagged donor cells that can be visualized in living tissue will help to determine whether non-resident cells integrating into the CNS microenvironment form functional connections characteristic of distinct neural cell types.

3.5.4.   *Bidirectional exchange between the immune system and the brain*

3.5.4.1. *Properties of adult neural stem cells*

Many tissues, including skin and liver, constantly repair themselves throughout life by replacing dead or dying cells with new cells derived from resident stem cells. The adult mammalian brain has long been thought to lack this regenerative capacity. However, the CNS contains stem and progenitor cells that generate new neurons and macroglia (oligodendrocytes and astrocytes) throughout embryonic and postnatal life [76, 90, 91,

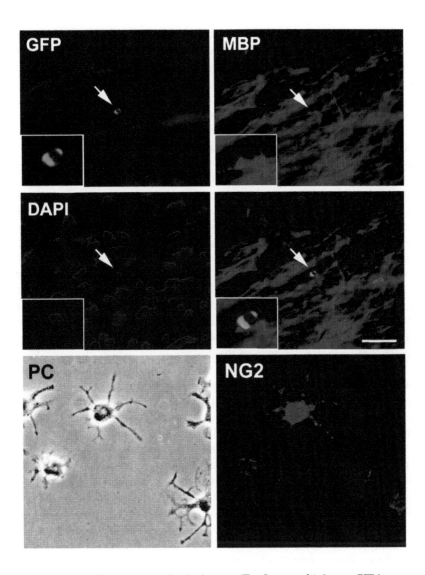

Figure 4.   Bone marrow cells can become oligodendrocytes. **(Top four panels)** shows a GFP immunoreactive oligodendrocyte (arrow) in the corpus callosum of a female mouse that received an intraventricular transplant of bone marrow stem cells 3 days after birth. GFP (green) and the oligodendrocyte-specific protein, MBP (red) were detected by immunofluorescence. DAPI staining (blue) was performed to identify the nucleus in individual cells. The triple overlay of GFP, MBP, and DAPI is also shown. The results suggested that bone marrow cells integrated into white matter tracts and expressed markers of mature oligodendrocytes. **(Bottom two panels)** Sca1 immunopositive hematopoietic stem cells were isolated from total bone marrow cells by magnetic separation, using the Miltenyi Biotech Multisort MicroBeads kit. Cells were grown in the presence of bFGF and epidermal growth factor. After 30 days, some NG2 immunopositive cells (red) were present that were morphologically similar (phase contrast micrograph) to pro-oligodendrocytes. Scale bar represents: 20 μm, top 4 panels; 13 μm, bottom 2 panels.

92, 93, 94, 95, 96, 97, 98, 99, 100]. Therefore, under the appropriate conditions, the brain is capable of self-repair. Over the past decade, researchers have focussed on identifying ways to mobilize stem cells to generate new functional neurons and glial cells after injury or disease.

3.5.4.2. *Sources of neural stem cells in the adult central nervous system*

Because of their apparent plastic nature, neural stem cells have become the focus of intensive research aimed at developing transplantation strategies to promote neural recovery in the diseased or injured nervous system [94, 101]. Several studies suggested that a single population of stem cells exists in the CNS [102, 103], whereas others demonstrated functional differences between stem cells [104, 105, 106]. Two populations of stem cells have been reported in the adult CNS; ependymal cells (a single layer of cells that line the ventricle; [98]) and those expressing GFAP in the subventricular zone [76]. Both classes gave rise to neurons, astrocytes, and oligodendrocytes [97, 107, 108].

Neurogenesis is localized to limited areas of the adult nervous system [97] and this process is believed to provide the CNS with a constant source of new neurons throughout ontogeny. In the rodent brain there are two well-characterized germinal centers, characterized by high-density cell division [83, 90, 92, 109, 110]. The first is the subventricular zone in the forebrain adjacent to the wall lining the lateral ventricle. New interneurons involved in sensing odors are formed here for the olfactory bulb. The second is the subgranular zone of the dendate gyrus, which gives rise to granule cell interneurons. When stem cells obtained from neurogenic regions were reintroduced back into the adult nervous system, they generated new neurons and glia that were characteristic of the endogenous population [107, 111].

More recently, the continuous formation of new cells was identified in the intact neocortex of adult mice [112] and primates [113]. Magavi and colleagues used an injury model, which involved synchronous apoptotic degeneration, to show that endogenous neural precursors differentiated into neocortical neurons. The process of differentiation occurred in a layer- and region-specific manner and resulted in neurons re-forming appropriate corticothalamic connections in regions of adult neocortex that did not normally undergo neurogenesis. Some of the newly formed cells in the neocortex expressed neuronal markers and extended processes to the original target sites, hinting at the possibility that these newly formed neurons established functional circuits. Whether these newborn neurons originated from the cortex, the SVZ, or some other source, like the peripheral circulation, is still unknown. However, these studies support the view that under the appropriate conditions, the adult brain is capable of regenerating in response to injury.

3.5.4.3. *Brain to blood: neural stem cell plasticity*

An increasing number of studies suggest that stem cells found in the nervous system are limited to neural fates because of the environmental limitations imposed upon them. For example, fetal stem cells grafted into the developing brain migrated in conjunction with host cells and differentiated into the appropriate cell types for that target region [69, 114, 115, 116]. In a provocative study, Bjornson and colleagues [117] demonstrated that genetically marked neural stem cells isolated from the ventricular zone of ROSA26 mice could become blood cells, a very different fate than their well-characterized neural derivatives. Isolated clones that normally gave rise to neurons, astrocytes, and oligodendrocytes were injected into the tail veins of irradiated mice. After 20 weeks, both the blood and bone marrow of recipients contained blood cells derived from the transplanted neural stem cell pool. These nervous system-derived cells transformed into a variety of hematopoietic cells, including cells of the myeloid and the lymphoid lineages, as well as more immature blood cells.

Adult neuronal stem cells, when co-cultured with totipotent embryonic stem cells, expressed genes characteristic of immature striated muscle, such as desmin and myosin heavy chain [118]. Although the differentiation of these adult cells into muscle was not complete, cultured CNS fetal stem cells could be pushed to become skeletal muscle [119]. Galli et al. [120] showed that acutely isolated and clonally derived adult neural stem cells from mice and humans, respectively, produced skeletal myotubes *in vitro* and following transplantation into adult mice.

4.   Routes of exchange between the CNS and immune system

The studies described herein suggest that there may be a continuous influx of cells from the bone marrow into the CNS and, simultaneously, from the brain into the peripheral circulation. Based on our analysis of the distribution of donor cells in transplant recipients, we propose that the circulating bone marrow progenitor cells reach the CNS through the choroid plexus, travel through the cerebrospinal fluid to cross the ependyma, and seed the brain parenchyma. The cerebral spinal fluid may provide the necessary environmental cues that initiate the differentiation of cells down one or more neural pathways. Magnetic resonance imaging is becoming an increasingly sensitive approach to follow the migration of transplanted cells within living organisms [121, 122] and may provide a powerful approach to identifying the entry routes of cells crossing different organ systems. Moreover, the use of cell specific promoters to drive the expression of fluorescent proteins or enzymes in specific cell types may be a useful approach to examining cellular differentiation in living tissue and allow for the electrophysiological characterization of cells. Molecular studies (e.g., microarray analysis and proteomics) of bone marrow and neural cells will help in the design of a panel of markers for each cell type that can be used to identify cells and help to better characterize the fate of bone marrow cells that enter the CNS. These approaches will also help to alleviate controversy over whether cells entering the brain from the periphery simply share antigenic properties in common with neural cells.

5.   Therapeutic potential of bone marrow stem cells for the treatment of CNS disease and injury

The ethical issues surrounding the use of human embryonic/fetal stem cells for research purposes is currently under hot debate and such discussions may impede our progress in using these cells to treat human disease. As a result, many researchers are beginning to investigate sources of adult stem cells and examine the potential for these cells in the treatment of various human diseases (Table 1). If adult bone marrow stem cells are as versatile as recent studies suggest, the therapeutic implications for CNS etiologies are significant. When designing transplantation strategies to treat neurological disease, donor cells should be 1) easily accessible, 2) immunologically inert, 3) capable of integrating into the nervous system and surviving long-term and, if necessary, 4) easily expanded *in vitro*. Bone marrow stromal stem cells exhibit all of these characteristics and are far more accessible than adult neural stem cells. Current efforts by independent laboratories are aimed at determining whether hematopoietic stem cells are similarly capable of differentiating into distinct neural cell types. However, it will be necessary to identify the most efficacious pharmacological treatments needed to safely enrich, instruct, or select for distinct CNS cell types.

It is likely that the influx of endogenous bone marrow stem cells from the periphery might be a contributing factor of the CNS manifestation of diseases that primarily originate in and affect the hematopoietic system (leukemia, AIDS etc). Such considerations must be addressed prior to using cells for autologous grafting. Moreover, the majority of stem cell plasticity-related studies have been done in mouse models. Since mouse stem cells differ from those of human origin, future efforts need to examine whether human bone marrow cells have a similar "neural" potential.

Other important issues include determining whether bone marrow-derived stem cells must pass through the general circulation or be exposed to the cerebrospinal fluid of the CNS before acquiring their neural potential. As some studies suggest, there may be a large number of pluripotent bone marrow-derived stem cells present in the adult mouse brain [58, 123]. If non-neural stem cells can migrate into the CNS, it will be useful to identify ways to mobilize these cells to regions of disease or injury CNS and encourage them to differentiate down specific lineages. It will be important to determine whether there are temporal limitations dictating how long it takes for peripherally-derived cells to seed the brain, when cells can cross the blood brain barrier, and the relative role of transient or chronic injury or disease. If recipient age is a limiting factor, then what are the optimal entry routes for transplanting cells?

6. Conclusions

Classically, studies have suggested that fetal stem cells are much more versatile in nature than those found in the adult mammal. However, many different sources of stem cells have been identified in adult animals and, under the appropriate conditions, these cells exhibit remarkable plasticity. One way to circumvent ethical issues surrounding the use of human embryonic/fetal tissue for the treatment of disease is to determine whether alternative sources of stem cells are equally competent (Table 1). Easily accessible sources of cells exhibiting a wide range of developmental potentials for future therapies may include those obtained from the adult bone marrow or cord blood [124]. However, it remains to be determined whether adult stem cells have the same broad potential as stem cells derived from embryos or fetal tissue, underscoring the need for scientists to carefully analyze and compare these different resources.

7. Acknowledgments

The authors would like to thank Drs. Rick Cohen and Steven Goldman for critical review of this chapter. KJC would like to dedicate this chapter to Drs. Lynn Hudson and John Kessler, who encourage and support progressive scientific ideas, and to Tyler, an inspiration.

8. References

1. Maves L., and Schubiger, G. (1998) Development 125, 115-124.
2. Matsuoka, S., Ebihara, Y., Xu, Mj, Ishii, T., Sugiyama, D., Yoshino, H., Ueda, T., Manabe, A., Tanaka, R., Ikeda, Y., Nakahata, T., and Tsuji, K. (2001) Blood 97, 419-425.
3. Ikuta, K., Kina, T., Macneil, I., Uchida, N., Peault, B., Chien, Y.H., and Weissman, I.L. (1990) Cell 62, 863-874.
4. Wood, W.G., Bunch, C., Kelly, S., Gunn, Y., and Breckon, G. (1985) Nature 313, 320-323.
5. Kantor, A.B., Stall, A.M., Adams, S., Herzenberg, L.A., and Herzenberg, L.A. (1992) Proc. Natl. Acad. Sci. USA 89, 3320-3324.
6. Michejda, M., Bellanti, J.A., Mazumder, A., Verma, U.N., and Wu, A.G. (1996) Fetal Diagn. Ther. 11, 373-382.
7. Qian, X., Shen, Q., Goderie, S.K., He, W., Capela, A., Davis, A.A., and Temple, S. (2000) Neuron 28, 69-80.
8. Notaro, R., Cimmino, A., Tabarini, D., Totoli, B., and Luzzatto, L. (1997) Proc. Natl. Acad. Sci. USA 94, 13782-13785.
9. Bodnar, A.G., Oullette, M., Frolkis, M., Holt, S.E., Chiu, C.P., Morin, G.B., Harley, C.B., Shay, J.W., Lichtsteiner, S., and Wright, W.E. (1998) Science 279, 349-352.
10. Reddel, R.R. (1998) Ann. N.Y. Acad. Sci. 20, 8-19.
11. Rudolph, K.L., Chang, S., Lee, H.W., Blasco, M., Gottlieb, G.J., Greider, C., and DePinho, R.A. (1999) Cell 96, 701-712.
12. Cheng, T., Rodrigues, N., Dombkowski, D., Stier, S. and Scadden, D.T. (2000) Nature Med. 6, 1235-1240.
13. Wognum, A.W., de Jong, M.O., and Wagemaker, G. (1996) Leuk Lymphoma 24, 11-25.
14. Qian, X., Davis, A.A., Goderie, S.K., and Temple, S. (1997) Neuron 18, 81-93.
15. Palm, K., Salin-Nordström, T., Levesque, M.F., Neuman, T. (2000) Mol. Brain Res. 78, 192-195.
16. Blau, H.M. and Baltimore, D. (1991) J. Cell Biol. 112, 781-783.
17. Wilmut, I., Schnieke, A.E., McWhir, J., Kind A.J., and Campbell, K.H. (1997) Nature 385, 810-813.
18. Civin, C.I., Trischmann, T., Kadan, N.S., Davis, J., Noga, S., Cohen, K., Duffy, B., Groenewegen, I., Wiley, J., Law, P., Hardwick, A., Oldham, F., and Gee, A. (1996) J. Clin. Oncol. 14, 2224-2233.
19. Weissman, I.L. (2000) Science 287, 1442-1446.
20. Friedenstein, A.J., Gorskaja, U, and Kulagina, N.N. (1976) Exp. Hematol. 4, 276-274.
21. Beresford, J.N., Bennett, J.H., Devlin, C., Leboy, P.S., and Owen, M.E. (1992) J. Cell Sci. 102, 341-351.
22. Pereira, R.F., Halford, K.W., O'Hara, M.D., Leeper, D.B., Sokolov, B.P., Pollard, M.D., Bagasra, O., and Prockop, D.J. (1995) Proc. Natl. Acad.. Sci..USA 92, 4857-4861.
23. Prockop, D.J. (1997) Science 276, 71-74.
24. Charbord, P., Lerat, H., Newton, I., Tamayo, E., Gown, A.M., Singer, J.W., and Herve, J.W. (1990) Exp Hematol 18, 276-282.

25. Ferrari, G. Cusella-De Angelis. G., Coletta. M., Paolucci, E., Stornaiuolo, A., Cossu, G., and Mavilio, F. (1998) Science 279, 1528-1530.

26. Bittner, R.E., Schofer, C., Weipoltshammer, K., Ivanova, S., Streubel, B., Hauser, E., Freilinger, M., Hoger, H., Elbe-Burger, A., and Wachtler, R. (1999) Anat. Embryol. (Berl.) 199, 391-396.

27. Gussoni, E., Soneoka, Y., Strickland, C.D., Buzney, E.A., Khan, M.K., Flint, A.F., Kunkel L.M., and Mulligan, R.C. (1999) Nature 401, 390-394.

28. Liechty, K.W., MacKenzie, T.C., Shaaban, A.F., Radu, A., Moseley, A.B., Beans, R., Marshak, D.R., and Flake, A.W. (2000) Nature Med. 6, 1282-1286.

29. Wakitani, S., Saito, T., and Caplan, A.J. (1995) Muscle Nerve 18:1417-1426.

30. Jackson, K.A., and Goodell, M.A. (1999) Proc. Natl. Acad. Sci. USA 96, 14482-14486.

31. Flamme, I., and Risau, W. (1992) Development 116, 435-439.

32. Folkman, J., and Shing, Y. (1992) J. Biol. Chem. 267, 10931-10934.

33. Frazier, O.H., Baldwin, R.T., Eskin, S.G., and Duncan, J.M. (1993) Texas Heart Inst. J. 20, 78-82.

34. Asahara, T., Murohara, T., Sullivan, A., Silver, M., van der Zee, R., Li, T., Witzenbichler, B., Schatteman, G., and Isner, J.M. (1997) Science 275, 964-967.

35. Civin, C.I., Strauss, L.C., Brovall, C., Fackler, M.J., Schwartz, J.F., and Shaper, J.H. (1984) J. Immunol. 133, 157-165.

36. Fina, L., Molgaard, H.V., Robertson, D., Bradley, N.J., Monoghan, P., Delia, D., Sutherland, D.R., Baker, M.A., and Greaves, M.F. (1990) Blood 75, 2417-2426.

37. de Vries, C., Escobedo, J.A., Ueno, H., Houck, K., Ferrara, N., and Williams, L.T. (1992) Science 255, 989-991.

38. Morrison, S.J., Uchida, N., and Weissman, I.L. (1995) Annu. Rev. Cell Dev. Biol. 11, 35-71.

39. Shalaby, F., Rossant, J., Yamaguchi, T.P., Gertsenstein, M., Wu, X.F., Breitman, M.L., and Schuh, A.C. (1995) Nature 376, 62-66.

40. Shi, Q., Rafii, S., Wu, M.H., Wijelath, E.S., Yu, C., Ishida, A., Fujita, Y., Kothari, S., Mohle, R., Sauvage, L.R., Moore, M.A., Storb, R.F. and Hammond, W.P. (1998) Blood 92, 362-367.

41. Kocher, A.A., Schuster, M.D., Szabolcs, M.J., Takuma, S., Burkhoff, D., Wang, J., Homma, S., Edwards, N.M., and Itescu, S. (2001) Nature Med. 7, 430-436.

42. Thorgeirsson, S.S., Evarts, R.P., Bisgaard, H.C., Fujio, K., and Hu, Z. (1993) Proc. Soc. Exp. Biol. Med. 204, 253-260.

43. Fausto, N. (1994) in: I.M. Arias, J.L. Boyer, N. Fausto, W.B. Jakoby, D.A. Schachter, D.S. Shafritz (Eds.), The Liver: Biology and Pathobiology, 3rd ed., Chapt. 78, Raven, New York.

44. Grisham, J.W. and Thorgeirsson, S.S. (1997), in: C.S. Potten (Ed.), Stem Cells, chapt. 8, Academic Press, San Diego, CA.

45. Fujio, K., Evarts, R.P., Hu, Z., Marsden, E.R., and Thorgeirsson, S.S. (1994) Lab. Invest. 70, 511-516.

46. Omori, N., Omori, M., Evarts, R.P., Teramoto, T., Miller, M.J., Hoang, T.N., and Thorgeirsson, S.S. (1997) Hepatology 26, 720-727.

47. Petersen, B.E., Goff, J.P., Greenberger, J.S., and Michalopoulos, G.K. (1998) Hepatology 27, 433-445.

48. Taniguchi, H., Toyoshima, T., Fukao, K., and Nakauchi, H. (1996) Nature Med. 2, 198-203.

49. Petersen, B.E., Bowen, W.C., Patrene, K.D., Mars, W.M., Sullivan, A.K., Murase, N., Boggs, S.S., Greenberger, J.S., and Goff, J.P. (1999) Science 284, 1168-1170.

50. Lagasse, E., Connors, H., Al-Dhalimy, M., Reitsma, M., Dohse, M., Osborne, L., Wang, X., Finegold, M., Weissman, I.L., and Grompe, M. (2000) Nature Med. 6, 1229-1234.

51. Grompe, M., al-Dhalimy, M., Finegold, M., Ou, C.N., Burlingame, T., Kennaway, N.G., and Soriano, P. (1993) Genes Dev 7, 2298-2307.

52. Overturf, K., Al-Dhalimy, M., Tanguay, R., Brantly, M., Ou, C.N., Finegold, M., and Grompe, M. (1996) Nat. Genet. 12, 266-273; erratum: 12, 458.
53. Hickey, W.F. (1999) Semin. Immunol. 11, 125-137.
54. Hickey, W.F., Hsu, B.L., and Kimura, H. (1991) J. Neurosci. Res. 28, 254-260.
55. Williams, K.C. and Hickey, W.F. (1995) Curr. Top. Microbiol. Immunol. 202, 221-245.
56. Knopf, P.M., Harling-Berg, C.J., Cserr, H.F., Basu, D., Sirulnick, E.J., Nolan, S.C., Park, J.T., Keir, G., Thompson, E.J., and Hickey, W.F. (1998) J. Immunol. 161, 692-701.
57. Theele, D.P. and Streit, W.J. (1993) Glia 7, 5-8.
58. Eglitis, M.A. and Mezey, E. (1997) Proc. Natl. Acad. Sci. USA 94, 4080-4085.
59. Suzuki, K. and Taniike, M. (1995) Micros. Res. Tech. 32, 204-214.
60. Taniike, M., Mohri, I., Eguchi, N., Irikura, D., Urade, Y., Okada, S., and Suzuki, K. (1999) J. Neuropathol. Exp. Neurol. 58, 644-653.
61. Wu, Y.-P.., Matsuda, J., Kubota, A., Suzuki, K., and Suzuki, K. (2000) J. Neuropath. Exptl. Neurol. 59, 628-639.
62. Wu, Y.-P., McMahon, E., Kraine, M.R., Tisch, R., Meyers, A., Frelinger, J., Matsushima, G.K., and Suzuki, K. (2000) Amer. J. Pathol. 156, 1849-1854.
63. Azizi, S.A., Stokes, D., Augelli, B.J., DiGirolamo, C., and Prockop, D.J. (1998) Proc. Natl. Acad. Sci. USA 95, 3908-3913.
64. Kopen, G.C., Prockop, D.J., and Phinney, D.G. (1999) Proc. Natl. Acad. Sci. USA 96, 10711-10716.
65. Mezey E., Chandross, K.J., Harta, G., Maki, R.A., and McKercher, S.R. (2000) Science 290, 1779-1782.
66. McKercher, S.R., Torbett, B.E., Anderson, K.L., Henkel, G.W., Vestal, D.J., Baribault, H., Klemsz, M., Feeney, A.J., Wu, G.E., Paige, C.J. and Maki, R.A. (1996) E.M.B.O. J. 15, 5647-5658.
67. Tondravi, M.M., McKercher, S.R., Anderson, K., Erdmann, J.M., Quiroz, M., Maki, R. and Teitelbaum, S.L. (1997) Nature 386, 81-84.
68. Eng, L.F., Vanderhaeghen, J.J., Bignami, A., and Gerstl, B. (1971) Brain Res. 28, 351-354.
69. Wichterle, H., Garcia-Verdugo, J.M., Herrera, D.G. and Alvarez-Buylla, A. (1999) Nat. Neurosci. 2, 461-466.
70. Mullen, R.J., Buck, C.R. and Smith, A.M. (1992) Development 116, 201-211.
71. Wolf, H.K., Buslei, R., Schmidt-Kastner, R., Schmidt-Kastner, P.K., Pietsch, T., Wiestler, O.D. and Bluhmke, I. (1996) J. Histochem. Cytochem. 44, 1167-1171.
72. Sarnat, H.B., Nochlin, D. and Born, D.E. (1998) Brain Dev. 20, 88-94.
73. Lendahl, U., Zimmerman, L.B., and McKay, R.D. (1990) Cell 60, 585-595.
74. Nishiyam, A., Lin, X.H., Giese, N., Heldin, C.H., and Stallcup, W.B. (1996) J. Neurosci. Res. 43, 299-314.
75. Sommer, I., and Schachner, M. (1980) Dev. Biol. 83, 311-327.
76. Doetsch, F., Caille, I., Lim, D.A., Garcia-Verdugo, J.M., and Alvarez-Buylla, A. (1999) Cell 97, 703-716.
77. Mezey, E., and Chandross, K.J. (2000) Eur. J. Pharmacol. 405, 297-302.
78. Okabe, M., Ikawa, M., Kominami, K., Nakanishi, Y., and Nishimune, Y. (1997) F.E.B.S. Lett. 407, 313-319.
79. Brazelton, T.R., Fabio, Rossi, F.M.V., Keshet, G.I., and Blau, H.M. (2000) Science 290, 1775-1779.
80. Ledbetter, J.A., and Herzenberg, L.A. (1979) Immunol. Rev. 47, 63-90.
81. Springer, T., Galfre, G., Secher, D.S., and Milstein, C. (1978) Eur. J. Immunol. 8, 539-551.
82. Akiyama, H. and McGeer, P.L. (1990) J. Neuroimmunol. 30, 81-93.
83. Lois, C. and Alvarez-Buylla, A. (1993) Proc. Natl. Acad. Sci. USA 90, 2074-2077.

84. Luskin, M.B. (1993) Neuron 11, 173-189.
85. Lois, C. and Alvarez-Buylla, A. (1994) Science 264, 1145-1148.
86. Sanchez-Ramos, J., Song, S., Cardozo-Pelaez, F., Hazzi, C., Stedeford, T., Willing, A., Freeman, T.B., Saporta, S., Janssen, W., Patel, N., Cooper, D.R. and Sanberg, P.R. (2000) Exp. Neurol. 164, 247-256.
87. Woodbury, D., Schwarz, E., Prockop, D.J. and Black, I. (2000) J. Neurosci. Res. 61, 364-370.
88. Mokry, J., and Nemecek, S. (1998) Acta. Medica. 41, 73-80.
89. Niki, T., Pekny, M., Hellemans, K., Bleser, P.D., Berg, K.V., Vaeyens, F., Quartier, E., Schuit, F., and Geerts, A. (1999) Hepatology 29, 520-527.
90. Goldman, S.A., and Nottebohm, F. (1983) Proc. Natl. Acad. Sci. USA 80, 2390-2394.
91. Reynolds, B.A. and Weiss, S. (1992) Science 255, 1707-1710.
92. Weiss, S., Reynolds, B.A., Vescovi, A.L., Morshead, C., Craig, C.G. and van der Kooy, D. (1996) Trends Neurosci. 19, 387-393.
93. Weiss, S., Dunne, C., Hewson, J., Wohl, C., Wheatley, M., Peterson, A.C. and Reynolds, B.A. (1996) J. Neurosci. 16, 7599-7609.
94. McKay, R. (1997) Science 276, 66-71.
95. Alvarez-Buylla, A., and Temple, S. (1998) J. Neurobiol. 36, 105-110.
96. Mujtaba, T., Mayer-Proschel, M., and Rao, M.S. (1998) Dev. Biol. 200, 1-15.
97. Weiss, S. and van der Kooy, D. (1998) J. Neurobiol. 36, 307-314.
98. Johansson, C.B., Momma, S., Clarke, D.L., Risling, M., Lendahl, U., and Frisen, J. (1999) Cell 96, 25-34.
99. Gage, F.H. (2000) Science 287, 1433-1438.
100. Roy, N.S., Wang, S., Jiang, L., Kang, J., Benraiss, A., Harrison-Restelli, C., Fraser, R.A., Couldwell, W.T., Kawaguchi, A., Okano, H., Nedergaard, M., and Goldman, S.A. (2000) Nature Med. 6, 271-277.
101. Bjorklund, A. and Svendsen, C. (1999) Nature 397, 569-570.
102. Gritti, A., Frolichsthal-Scoeller, P., Galli, R., Parati, E.A., Cova, L., Pagano, S.F., Bjornson, C.R., and Vescovi, A.L. (1999) J. Neurosci. 19, 3287-3297.
103. Kukekov, V.G., Laywell, E.D., Suslov, O., Davies, K., Scheffler, B., Thomas, L.B., O'Brien, T.F., Kusakabe, M., and Steindler, D.A. (1999) Exp. Neurol. 156, 333-344.
104. Campbell, K., Olsson, M., Bjorklund, A. (1995) Neuron 15, 1259-1273.
105. Tropepe, V., Sibilia, M., Ciruna, B.G., Rossant, J., Wagner, E.F., van der Kooy, D. (1999) Dev. Biol. 208, 166-188.
106. Yaworsky, P.J. and Kappen, C. (1999) Dev. Biol. 205, 309-321.
107. Gage, F.H., Coates, P.W., Palmer, T.D., Kuhn, H.G., Fisher, L.J., Suhonen, J.O., Peterson, D.A., Suhr S.T., and Ray, J. (1995) Proc. Natl. Acad. Sci. USA 92, 11879-11883.
108. Zhang, S.C., and Duncan, I.D. (1999) Proc. Natl. Acad. Sci. USA 96, 4089-4094.
109. Morshead, C.M., Reynolds, B.A., Craig, C.G., McBurney, M.W., Staines, W.A., Morassutti, D., Weiss, S. and van der Kooy, D. (1994) Neuron 13, 1071-1082.
110. Luskin, M.B. (1998) J. Neurobiol. 36, 221-233.
111. Suhonen, J.O., Peterson, D.A., Ray, J. and Gage, F.H. (1996) Nature 383, 624-627.
112. Magavi, S.S., Leavitt, B.R., and Macklis, J.D. (2000) Nature 405:951-955.
113. Gould, E., Reeves, A.J., Graziano, M.S.A., and Gross, C.G. (1999) Science 286:548-552.
114. Brustle, O., Spiro, A.C., Karram, K., Choudhary, K., Okabe, S., and McKay, R.D. (1997) Proc. Natl. Acad. Sci. USA 94, 14809-14814.
115. Brustle, O., Choudhary, K., Karram, K., Huttner, A., Murray, K., Dubois-Dalcq, M., and McKay, R.D. (1998) Nat. Biotechnol. 16, 1040-1044.

116. Olsson, M., Bjerregaard, K., Winkler, C., Gates, M., Bjorklund, A. and Campbell, K. (1998) Eur. J. Neurosci. 10, 71-85.

117. Bjornson, C.R., Rietze, R.L., Reynolds, B.A., Magli, M.C., and Vescovi, A.L. (1999) Science 283, 534-537.

118. Clarke, D.L., Johansson, C.B., Wilbertz, J., Veress, B., Nilsson, E., Karlström, H., Lendahl, U., and Frisén, J. (2000) Science 288, 1660-1663.

119. Valtz, N., Hayes, T., Norregaard, T., Liu, S. and McKay, R. (1991) New Biol. 3, 364-371.

120. Galli, R., Borello, U., Gritti, A., Minasi, M.G., Bjornson, C., Coletta, M., Mora, M., De Angelis, M.G.C., Fiocco, R., Cossu, G., and Vescovi, A.L. (2000) Nat. Neurosci. 3, 986-991.

121. Bulte, J.W., Zhang, S.-C., van Gelderen, P., Herynek, V., Jordan, E.K., Duncan, I.D., and Frank, J.A. (1999) Proc. Natl. Acad. Sci. USA 96:15256-15261.

122. Franklin, R.J.M., Blaschuk, K.L., Bearchell, M.C., Prestoz, L.L.C., Setzu, A., Brindle, K.M., and Ffrench-Constant, C. (1999) Neuroreport 10:3961-3965.

123. Barlett, P.F. (1982) Proc. Natl. Acad. Sci. USA 79, 2722-2725.

124. Williams, D.A., and Moritz, T. (1994) Blood Cells 20, 504-515.

125. Tropepe, V., Coles, B.L., Chiasson, B.J., Horsford, D.J., Elia, A.J., McInnes, R.R., and van der Kooy, D. (2000) Science 287, 2032-2036.

126. Pellegrini, G., Golisano, O., Paterna, P., Lambiase, A., Bonini, S., Rama, P., De Luca, M. (1999) J. Cell Biol. 145, 769-782.

127. Taylor, G., Lehrer, M.S., Jensen, P.J., Sun, T.T., and Lavker, R.M. (2000) Cell 102, 451-461.

128. Bjerknes, M., and Cheng, H. (1981) Am. J. Anat. 160, 65-75.

129. Cohn, S.M., Roth, K.A., Birkenmeier, E.H., and Gordon, J.I. (1991) Proc. Natl. Acad. Sci. USA 88, 1034-1038.

130. Theise, N.D., Badve, S., Saxena, R., Henegariu, O., Sell, S., Crawford, J.M., and Krause, D.S. (2000) Hepatology 31, 235-240.

CHAPTER 5

AGING AND NEURAL STEM CELLS

JINGLI CAI and MAHENDRA RAO

Table of contents

1.   Introduction

Organisms appear to have a limited lifespan with a decline of peak function over time that ultimately leads to death. Nothing has been demonstrated to slow or reverse the primary aging process in humans [1]. This has led to the concept of an inherent limit in lifespan (discussed elsewhere in this book). However the median age of death in the human population is much shorter than the postulated maximum age of survival raising the possibility than additional longevity can be achieved by preventing the secondary effects of aging. Indeed all the factors that are known to affect longevity do so by modifying or delaying the development of age related disorders — a part of secondary aging. Thus much of the work in aging is aimed at developing preventive strategies against secondary aging: strategies aimed at maintaining health and functional capacity and rectangularizing, rather than extending, the survival curve.

Stem Cells: A Cellular Fountain of Youth.   Ed. by Mark P. Mattson and Gary Van Zant. 97 — 116

The progressive decline in function due to age related disorders (secondary aging) is nowhere more apparent than in the nervous system where striking changes in morphology, connectivity and function have been documented [2]. What has been noted repeatedly is that the adult CNS lacks effective strategies for functional replacement [3]. Most CNS repair strategies that do exist involve compensatory mechanisms (summarized in Table 1), take advantage of the inherent redundancy in neural systems, and do not involve generation of new cells. However, in selected regions actual replacement of neurons has been documented. In particular, cellular replacement has been shown to occur in the olfactory bulb and in the hippocampus. It is important to point out that, while a failure to see neuronal replacement is the norm in most brain regions, astrocyte proliferation and oligodendrocyte replacement is seen throughout the neuraxis. Division of astrocytes has been documented in the uninjured brain with a massive increase after ischemic or traumatic insults. Oligodendrocytes themselves do not re-enter the cell cycle but the presence of oligodendrocyte precursors, which proliferate and remyelinate after demyelination episodes, has been described.

The nervous system may however have a much higher potential for regeneration or repair than previously thought. Stem cells and precursor cells have been identified in the fetal brain [4-15] as well as in the adult human subependymal zone and hippocampus [16-17] and stem cell lines propogated by genetic or epigenetic means have been obtained [11-13]. The number of such cells has been estimated and their self renewal ability may be sufficient to serve as a source for repair either by mobilizing endogenous stem cells and precursor cells or by transplanting such cells at an appropriate time.

The discovery that stem cells exist and can be utilized for repair has raised the possibility that stem cell therapy may be useful in retarding or reversing some of the effects of secondary aging. While there is increasing interest in stem cells, the data available for adult stem cells and their role in aging remains limited. In this chapter we have therefore presented current information about human stem cells and precursor cells, properties of adult stem cells that may allow them to be used as successful vehicles for replacement therapy and described potential therapeutic strategies.

Table 1.   Mechanisms for nervous system repair.

Functional recovery after neuronal damage

Restitution due to local repair	eg: Restoration of vision after cortical stomata
Restitution due to redundancy	eg: Restoration of function after strictly unilateral damage
Restitution by substitution	eg: Recovery of some visual function after damage to primary visual areas
Restitution by compensation	eg: Recovery after unilateral labryinthine damage
Recovery by cellular replacement	eg: Olfactory bulb regeneration

Several mechanisms operate to restore function after neuronal damage in the adult brain. It should be noted that functionally appropriate cellular replacement is normally seen in the olfactory system. Glial replacement may however occur after demyelination lesions or ischemic damage (see text for details).

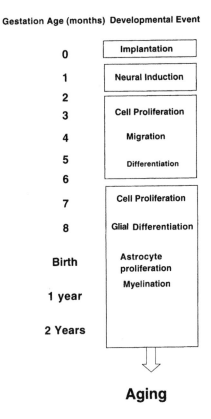

Gestation Age (months) Developmental Event

Gestation Age (months)	Developmental Event
0	Implantation
1	Neural Induction
2	
3	Cell Proliferation
4	Migration
5	Differentiation
6	
7	Cell Proliferation
8	Glial Differentiation
Birth	Astrocyte proliferation
	Myelination
1 year	
2 Years	

Aging

Figure 1. Stages of development. Development of the nervous system occurs in multiple steps that extend through a significant period of gestation. A timetable of development modified from [18] is shown. Note that glial differentiation is temporally dissociated from neuronal differentiation and extends postnatally. Not shown is myelination after oligodendrocytes have differentiated. Myelination is tract specific and extends through childhood with some tracts not fully myelinated until age twelve.

2.   Human neural stem and precursor cells

Neural development can be divided into several developmental stages that occur at specific gestation ages (Figure 1). Neural induction is the process by which the neural tube, neural crest and placodal precursors are generated. This occurs between 3 and 5 weeks of gestation (reviewed in 18). Cells present at this stage in the neural tube are termed neuroepithelial cells and do not express markers of differentiated neurons, astrocytes and oligodendrocytes. Induction is followed by a period of neuroepithelial cell proliferation, and patterning which extends over several additional weeks with neuronal generation seen through 20-22 weeks of gestation. Proliferation and patterning overlap a period of neuronal migration and aggregation (Table 2). This period is followed by cytodifferentiation, formation of specific connections, cell death and synapse elimination. Glial differentiation overlaps neuronal generation but its onset is somewhat delayed relative to neurogenesis (reviewed in [19]). Astrocyte and oligodendrocyte production reaches a peak postnatally and likely continues throughout life. Lineage relationships between neuroepithelial cells and differentiated neural cells have been established in rodent tissue and a possible lineage diagram is shown (Figure 2). The lineage relationships between human multipotent stem cells, restricted precursors and differentiated cells have not been described but are inferred based on the sequential appearance of these precursors during development. Human neuroepithelial stem cells, neuronal precursors and glial precursors however, have been identified.

Table 2.　　Stages of neuronal maturation

Stages in neuronal development
Induction
Proliferation
Migration
Aggregation
Cytodifferentiation
Axonal outgrowth and synaptogenesis
Cell death
Synapse elimination

Neuronal development has been divided into multiple stages that occur over a prolonged period of several months during gestation.

While the exact lineage relationships are still being worked out, there is an emerging consensus that neural development is characterized by a progressive restriction in developmental fate (Figure 2). Seminal work by a number of laboratories has lead to rapid advances in our understanding of phenotypic specification. A complex terminology that is still evolving has developed to distinguish between these multiple stages of differentiation. For the purposes of this review, we will define a stem cell as any cell that is capable of self-renewal and can undergo asymmetric division to generate differentiated cells. Cell which have a wide repertoire of differentiation, including most cells in a particular tissue or organ, will be termed multipotent. The terms "bipotent" and "unipotent" will be used to describe stem cells that generate two or one class of differentiated progeny. Progenitor/blast and transit amplifying cells are, in our opinion, equivalent terms and denote cells with a limited self renewing capacity and a relatively restricted repertoire of differentiation [20]. Neuron restricted precursors (NRP) and glial restricted precursors (GRP) are restricted precursors that generate neurons and glia, respectively.

Using this definition, stem cells and precursor cells have been identified in both adult and fetal tissue from multiple regions. Characteristics of these cells in the adult appear

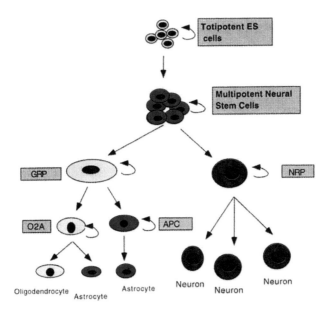

Figure 2. Lineage relationships between stem cells and precursor cells. The relationship between ES cells, NEP cells and intermediate precursors are shown. Neural stem cells (NSC's), glial restricted precursors (GRP's) and neuronal restricted precursors (NRP's) have all been shown to differentiate from NEP cells and to be restricted in their developmental potential, GRP's, O2A's and NRP's are self-renewing precursor cells that can be distinguished from each other and from NEP cells by their differentiation potential, cell surface markers, gene expression and cytokine response. Note that other possible restricted precursors could exist and additional stages in the process of differentiation may be required.

to be somewhat different than those of cells isolated from fetal tissue and these cells are generally distinguished by a prefix adult or fetal to the term stem cell. In addition to stem cells in the CNS, stem cells in other neural regions have been identified. These cells are similar in that they exhibit self-renewal capability and the ability to differentiate into multiple kinds of neurons and glial cells but they differ from each other in their relative ability to generate specific classes of cells. For example, retinal stem cells generate retinal derivatives far more readily than forebrain or spinal cord derived stem cells. Likewise mesenchephalic stem cells appear to generate dopaminergic neurons far more easily than any other precursor cell population [21]. Some of the different kinds of stem cells and precursor cells isolated from human tissue are listed (Table 3).

In addition to multipotent stem cells, restricted precursors have also been identified in human tissue. Neuronal precursors have been isolated from the developing human brain by fluorescence sorting for GFP expressing cells using constructs where GFP expression is under the control of the early promoter of Talpha-1 tubulin [17]. This promoter is uniquely expressed by neurons at the earliest stage of neuronal differentiation. Goldman and colleagues have reported the successful use of these neuronal promoter-targeted

Table 3.   Stem and precursor cells in human tissue.

Multipotential Stem cells

Fetal human stem cells	eg: Vescovi et al., 1999
Adult human stem cells	eg: Carpenter et al., 1999
Retinal stem cells	eg: Ezeonu et al., 1999
Neural crest stem cell	eg: ------------
Placodal stem cell	eg: Pagano et al., 2000

Neuron restricted precursors

N-CAM immunoreactive HNRP cells	eg: Piper et al., 2001#
Alpha-I tubulin expressing fetal HNRP cells	eg: Roy et al. et al., 2001
Alpha-I tubulin expressing adult HNRP cells	eg: Roy et al., 2001

Glial restricted precursors

Oligodendrocyte precursors fetal	eg: Zhang et al., 2000
Oligodendrocyte precursors adult	eg: Scolding 1998
Bipotential oligodendrocyte-astrocyte precursors fetal	eg: Mayer-Proschel et al. *
Astrocyte precursor cells	eg: Loo et al., 1991

Some of the stem cells and precursor cells identified are listed. A neural crest stem cell has been described but remains unpublished. Note that pineal stem cells have not as yet been isolated from human tissue. Only a single reference is provided due to space limitations. It is important to emphasize that several investigators have identified some or all of these cell types. * personal communication. # Manuscript in review.

selection techniques by either transfection or adenoviral infection to obtain neuronal precursors, from adult human hippocampus. A subset of the neuronal precursors obtained by this selection technique divide *in vitro* and most mature to form physiologically active neurons [17]. Using the same technique, neuronal precursors have also been isolated from the adult ventricular zone [22] and from fetal tissue. Rao and colleagues have likewise used antibodies to polysialated NCAM, which at early developmental stages is expressed by neurons and their precursors, to isolate human neuronal precursor cells from fetal tissue [23, 24]. Comparison of the expression pattern between the fetal and adult neuronal precursors suggests that these populations may differ in their self renewal and differentiation potential though detailed comparisons are necessary to confirm this preliminary observation.

Oligodendrocytes, astrocytes and glial precursors have been identified from fetal tissue and adult brains (reviewed in [19]). Mayer-Proschel and colleagues have described the presence of an oligodendrocyte-astrocyte precursor similar to the GRP cell [25] present *in vivo* (Mayer-Proschel, University of Rochester, personal communication). Using a strategy similar to that used for identifying neuronal precursors, Goldman and colleagues used the CNP promoter to select for oligodendrocytes and oligodendrocyte precursors [26]. Scolding and colleagues [27] used adult human tissue to demonstrate the existence of an A2B5 immunoreactive glial progenitor that appeared similar to the O2A cell described in rodent tissue. Proliferative adult human oligodendrocyte precursors have been isolated using markers from adult human white matter [27-30]. Similarly, Quinn and colleagues [14] have described what appears to be astrocyte restricted precursor cells. These astrocyte-restricted precursors were obtained by sequentially passaging multipotent stem cells from cultured human spinal cord tissue. Astrocyte precursors have not been directly isolated from human tissue though at least two classes of astrocyte precursors [31, 32] have been isolated from rodent tissue. It is important to emphasize that distinguishing astrocyte precursors from astrocytes is difficult. Astrocytes, while normally quiescent in vivo, are capable of reentering the cell cycle and can be passaged almost indefinitely. Indeed, recent data suggest that glial cells and their precursors express high telomerase levels [33, 34] and may be spontaneously immortal. Furthermore, some evidence has suggested that at least some astrocytes maybe capable of dedifferentiating in culture and generating neurons [35, 36]. It is unclear if human glial precursors will have similar properties but given the overall similarity in the properties of human and rodent precursors, it may be likely.

Thus human fetal and adult neural tissue contains multiple classes of dividing cells. These include multipotent stem cells, more restricted precursor cells and cells such as astrocytes which while quiescent are capable of re-entering the cell cycle. These cells respond in times of injury and to external cues to proliferate and should in principle be able to provide functional cellular replacement. However, as discussed above, most repairs in the brain involve compensatory mechanisms rather than cellular replacement. An important question in the stem cell field that is the subject of intense is to determining what factors modulate the properties of stem cells in the adult and aging brain. In the next section we summarize the limited information available on this subject. We discuss a generic model of stem cell behavior and how normal and abnormal events may alter stem cell behavior and contribute to unsuccessful aging. More detailed discussion of the role of stem cells in specific disorders are presented elsewhere in this book.

3.   Stem cells in the adult and aging brain

The normal fate of the stem cell in development is summarized in Figure 3. A stem cell may remain quiescent and simply not enter the cell cycle. Alternatively, it may undergo apoptosis and not contribute to further development. Stem cells may undergo symmetric cell divisions to self renew or undergo terminal differentiation, or they may undergo asymmetric cell divisions to generate differentiated progeny as well as maintain a pool of stem cells. Finally, a stem cell may transit to another kind of stem cell (e.g. adult stem cell) or transdifferentiate to acquire a distinct set of properties. The total number of cells present in the nervous system, or any tissue for that matter, represents a

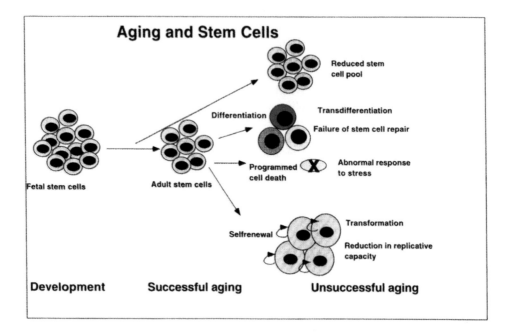

Figure 3.   Stem cells and Aging. Stem cells persist into adulthood and can differentiate, die, transform or transdifferentiate. Appropriate response to environmental signals is seen in successful aging while aberrant response either due to abnormal stem cell number or inappropriate response to environmental signals will lead to an inappropriate stem cell response to stress and lead to unsuccessful aging.

dynamic balance between self-renewal, differentiation and death. This dynamic balance is modulated by excitatory and inhibitory signals to meet the demands of the tissue to which stem cells may supply newly differentiated cells. The final decision of a cell to self renew, differentiate or remain quiescent is dependent on an integration of multiple signaling pathways and at each instant will depend on cell density, its energetic metabolism, ligand availability, type and levels of receptor expression, and downstream cross-talks between distinct signaling pathways (reviewed in [37]).

The importance of adult stem cells in normal homeostasis and aging is being recognized. It is becoming clear that several molecules that were thought to modulate aging also regulate stem cell renewal, proliferation and survival. Successful aging may be modified because abnormalities in early development may alter the pool of stem cells. Stem cells themselves may undergo changes during the aging process. The process of replication throughout life may have specific consequences that may contribute to the aging process. Failure to differentiate appropriately or adequately in the aging brain may lead to unsuccessful aging. Loss of appropriate signals or failure of precursor cells to respond to such signals may likewise lead to unsuccessful aging (Figure 3).

There are two important points to be made about the consequences of aging and stem cells. Stem cells may respond to the same 'aging' factors that alter somatic cell responses (Table 4). However, stem cells appear to be more susceptible to stress, damage, and free radical production than other somatic cells. Further, even damage during early development may reduce the pool of precursors with which one enters adulthood. The pool of cells, while reduced, may be sufficient to meet the homeostatic needs of the young adult but may be insufficient to meet the stress of aging leading to an acceleration of the degenerative process of aging. For example, Lemaire et al. showed that prenatal stress will alter neurogenesis in the adult and that this alteration in neurogenesis may be associated with deficits in learning and memory [38].

Table 4. Aging processes that affect stem cells.

Aging in somatic and stem cells
Telomerase levels
Protein synthesis and degradation
Mitochondrial instability and mutations
Free radical scavenging
Generation of free radicals
accumulation of somatic mutations

Several mechanisms may operate to reduce stem cell activity during the process of aging. These mechanisms alter somatic cell response also but the consequences of alterations in stem cell behavior will be greater especially in tissue with high stem cell activity. It is important to emphasize that combinatorial effects are often more than additive and that the stochastic nature of these events may lead to clones of functionally distinct stem cells

A second point worth emphasizing is that a reduction in stem cell renewal as a consequence of aging will have more global consequences on tissue maintenance and overall homeostasis than aging of other somatic cells. Figure 3 summarizes the difference between successful and unsuccessful aging of stem cells. This effect of stem cell aging is likely to be more important in tissues where self-renewal is rapid and ongoing. Examples of such tissue would be the hematopoietic system, intestinal epithelium and skin. In the nervous system the consequences of stem cell aging are likely to be more restricted. Two regions where ongoing neurogenesis is observed are the hippocampus and the olfactory bulb. It is therefore likely that if stem cell alteration during aging was functionally important, then changes should be seen in these regions first. It is intriguing to note that many of the degenerative changes seen in aging affect memory and learning functions attributed to the hippocampus. In the next section we will highlight some of the issues related to studying stem cells in the adult brain.

4.   Age related changes in stem cells

Somatic cells show characteristic changes during the process of aging. These include a reduction in the ability to replicate (in dividing populations), the accumulation of somatic mutations, the inability to respond to extracellular cues, alteration in energy and protein metabolism, increased production of free radicals, and shortening of telomere ends (Table 4). Neural stem cells like any other somatic cell, may undergo these age related changes as well, though the data available for stem cells in general and neural stem cells in particular is quite limited.

The ability of hematopoietic stem cells to self renew, while more than adequate for the lifespan of the individual, declines with aging. A measurable difference in the response to stress from young versus aged hematopoietic precursors has been documented [39]. This ability depends on maintaining the pool of proliferating precursors that are sufficient to meet the needs of the aging organism. Intestinal crypt cells which are another rapidly dividing stem cell population also show a difference in the degree of apoptosis in response to stress or ionizing radiation [40]. Aging cells appear far more susceptible than young stem cells. The ability of neural precursor cells to self renew or differentiate can be altered by modulating free radical scavenging [41]. Neural stem cells show a decline in telomerase activity after culture [42] and levels of telomerase activity are lower in adult brain cells than in the embryo [43, 44] suggesting that some of the mechanisms implicated in the aging of somatic cells likely work in a similar fashion in the adult neural stem cell populations.

5.   Identifying stem cells in the aging brain

Successful aging will involve maintaining stem cell homeostasis and its appropriate integration to normally distributed external differentiation, migration and proliferation signals. Modulating these parameters by caloric restriction [45], mild stress [46], hormonal therapy [47, 48] and free radical scavenging [49], therapies that have been shown to work in somatic cells are likely to work in stem cells. Indeed, it was recently shown that dietary restriction increases the survival of newly-generated neural cells in the hippocampus of adult rats [50]. What is not known however is how effective these therapies will be and what the long-term consequences of these treatments will be on stem cell populations. Performing these experiments and assessing their effects on stem cell populations have been difficult as no specific markers for stem cells exist and even the location of stem cells in the adult is in doubt. Almost all experiments that study multipotent stem cells in the nervous system rely on functional criteria (the ability to self renew, divide and differentiate into multiple phenotypes) as a post-hoc criteria for identifying stem cells or on the absence of markers characteristic of differentiated cells. Using these criteria to localize stem cells or to study their behavior in aging animals *in vivo* is almost impossible. It is therefore not surprising that controversy exists as to the number of stem cells, their location and their properties. Some investigators have argued that stem cells in the adult are located in the subependymal zone [51], while others have suggested that the ependymal cell is the adult stem cells [9, 10]. Still others have suggested that the cortex and white matter contain quiescent stem cells that can be activated by appropriate stimuli [9, 52]. Substantial efforts have been made to identify markers that would prospectively and specifically identify stem cells in the adult brain. These experiments have yielded few useful markers. For example, we identified FGFR4 as a marker for stem cells in fetal tissue [53] but found that this marker is not detectable on adult stem cells. Recently Uchida and colleagues have shown that ACS133 labels a small proportion of human precursor cells [54]. However, this marker also labels intermediate precursors and astrocytes and is therefore not an optimal marker. Until these technical problems are solved, studying the role of stem cells in aging is fraught with difficulty and the initial focus of most laboratories is to localize stem cells and identify markers that will permit one to study the effects of factors that modulate aging.

6.   Differentiation

While a detailed discussion of factors involved in the process of differentiation is outside the scope of this review, it is important to note that differentiation is a multistep event under the control of cell autonomous and cell non-autonomous factors. Effects at any stage in the differentiation process will alter the number and phenotypes of the cells generated. The specific alteration observed in the aging brain would depend on the stage of differentiation affected. The earlier the effect of any particular agent, the larger the potential downstream consequences are and, paradoxically, a greater probability of compensation. If for example, stem cell numbers were reduced in the aged brain, then the number of dividing cells seen would be lower. If on the other hand response to stress were altered, then the increase of dividing cells after a particular environmental stress would be altered. If differentiation was affected, the number of cells that would integrate would be reduced while the number born would remain unchanged. If integration were affected, then perhaps there would be an increase in the number of aberrantly positioned cells. These sorts of experiments have only recently been initiated and most of the available data relate to the olfactory bulb or the hippocampus. We will review some of the findings below, but the reader is urged to exercise caution before generalizing from these findings.

In experiments where the integration of stem cell derived neurons and their persistence was analyzed, several investigators showed that newly born neurons integrate and send axons to appropriate sites [55]. The newly born cells were more resistant to adrenalectomy than older cells [56]. Not all stem cell derived-neurons make appropriate connections and some cells are present in ectopic locations with abnormal connections [57]. However, despite clear cut evidence for normal differentiation and normal connectivity in the adult hippocampus from stem cells, detailed analysis of overall cell number had suggested a decline in aging.

The possibility therefore remained that this decline was due to a reduction in stem cell capacity or reduced environmental signaling. McKay and colleagues [58] have suggested that the decline in hippocampal neurogenesis (reviewed in [59]) is not due to a reduction in stem cell number or a reduction in differentiation ability. Rather they have suggested that it is the elevated glucocorticoid production that is responsible for inhibiting stem cell proliferation and that a reduction in corticosteroid level will abrogate the age related decline in hippocampal neurogenesis. Kempermann and colleagues have suggested that the potential to generate new neurons is higher than the levels of ongoing neurogenesis [60] and that increased neurogenesis can be obtained by exposing animals to an enriched environment or by exercise. Likewise Dash and colleagues

[61] have shown that neurogenesis increases after loss of granule cells suggesting that the replicative capacity of stem cells is high and that these cells can respond to stress. Moreover, dietary restriction, a manipulation known to retard aging, increases the survival of newly generated neural cells in the hippocampus [50]. While detailed comparisons between adult and aged animals have not been performed, the available data does suggest that stem cells persist and can respond to environmental signals and that the latent potential is sufficient to meet the demands of the tissue.

The overall evidence suggests that the decline in granule cell number may be due to a reduction in environmental signals, and that modulating extrinsic signals may be sufficient to restore the age related decline (reviewed in [62]). In addition to steroids (see above), alterations in thyroid hormone levels [63], overall level of activity [64] and exposure to stress [38] can alter the number of newly born granule cell neurons

In the olfactory bulb the data, while more limited, suggest that neurogenesis is ongoing and that a balance between differentiation and apoptosis is maintained. Modulating extracellular signals can alter the number of newly born cells [65] and estrogen can alter the number of dividing cells [66]. Growth factor infusions can increase the proliferation of precursor cells [67] and ablation of the stem cell pool by radioactive thymidine can be replenished by stem cell proliferation [68]. Modulating stem cell differentiation in other brain regions by environmental manipulation has not been seen though target ablation has suggested that stem cells may respond to specific kinds for environmental changes [69].

In many brain regions it appears likely that specific inhibitory influences prevent stem cells from differentiating. This is best exemplified by the work of Dr. Gage and colleagues [70]. These investigators showed that stem cells were present in the adult spinal cord and that these cells could differentiate into neurons in a region of ongoing neurogenesis (hippocampus) but not when transplanted back into the spinal cord. Thus, an important area of investigation is determining the environmental signals that regulate stem cell differentiation, their distribution and age related changes.

A final point to note about stem cells and differentiation is that there appears to be regional specificity to stem cells and that this specificity develops early in embryonic development (reviewed in [71]). Aging may not affect all stem cell populations equally and differing responses to the same agent may be seen in different brain regions. Indeed, response to thyroid hormone or steroids is different between cerebellum and hippocampus or hippocampus and olfactory bulb. As noted above, caution needs to be exercised before generalizing from one region to another.

7.   Transformation

Aged tissue in many mammalian species including humans [72] exhibits a mix of atrophy, hyperplasia and benign neoplasia, increasing malignant neoplasia. Neoplastic transformation therefore may be considered one form of unsuccessful aging. An inevitable aspect of aging is the increasing probability of neoplastic transformation. Indeed, the incidence of many cancers increases rapidly in the aging population. It is generally thought that cancer is a multi-hit event where dividing cells accumulate replication errors. The rate of such error accumulation depends on rates of cell division, defects in error correction machinery or presence of mutational agents. Mutations that affect normal cell cycle regulatory events then lead to uncontrolled proliferation and subsequent tumor formation. The rate of tumor formation increases with age and in general, tumors arise in dividing cell populations. Thus tumors of neuronal origin are rare or absent in the adult system while tumors of glial origin are relatively common. Cerebellar tumors of neuronal origin are seen at early postnatal periods when cerebellar granule cell proliferation is common.

Why tumors should increase in aging tissue is open to speculation. Like all somatic cells, perhaps replicative error accumulation is the underlying cause. Another possibility is asynchronous clonal attenuation of cell types that participate in maintaining proliferative homeostasis [72, 73]. Thus, one important intersection of studies in aging and in stem cell biology is the understanding of how stem cell division is regulated and how it can be manipulated in dividing cells in the adult.

8.   Stem cells as therapy in the adult and aging brain

It is clear from examining the aging brain that many diseases include massive cellular loss that is not adequately restored. Parkinson's, Huntington's and Alzheimers disease are all associated with regional and specific losses of cells [74]. In all cases appropriate functional replacement does not appear to occur despite the existence of large numbers of stem cells in the adult brain. Understanding stem cell biology therefore offers the potential of therapeutic intervention. Several strategies can be envisaged and these are summarized in Table 5 and discussed below.

Table 5.   Therapeutic uses of stem cells.

Potential uses of stem cells in aging
Mobilizing endogenous stem cells
Hormones, low stress, increased activity, target manipulation
Transplanting dissociated cells for replacement
Multipotent stem cells
More restricted precursors
Differentiated cells
Defined mixtures of cells
Transplanting dissociated cells for drug or protein delivery
Gene discovery
Drug screening
Bioengineering devices

Several potential strategies that utilize stem cells and precursor cells for therapy are listed.

9.   Mobilizing endogenous stem cells

Two important pieces of data have suggested that mobilizing endogenous stem cells may be a viable approach to therapy. First, a huge body of evidence suggests that stem cells are present throughout the brain (reviewed in [71]). Second, stem cells that respond to extrinsic cues in the adult and differentiated cells appear to be able to follow available cues to reach appropriate sites of injury (reviewed in [62]). While little data is available in humans, there is excitement of this as potential therapy based on results from rodent models. The initial experiments were focused on the hippocampus and the olfactory bulb where endogenous neurogenesis has been shown to occur. In these experiments investigators were able to show that growth factor infusion, exercise, hormone therapy, diet and manipulating the target could all be used to alter the rate of stem cell division, the survival of differentiated cells and the phenotype of the neurons that needed to be generated [50, 56, 57, 58, 61, 64, 65, 66]. More recently Dr. Macklis and colleagues have extended these observation to the cortex [69]. These investigators have shown that target ablation can alter the proliferation of cortical stem cells such that layer specific differentiated neurons are generated. These neurons integrate into the parenchyma in a site-specific fashion, receive appropriate synapses and connect to appropriate targets in the thalamus or the other cortex. Not all regions of the brain appear to be equally responsive. In the spinal cord and brainstem, lesions do not appear to stimulate neurogenesis and even transplanted stem cells fail to differentiate into neurons [70].

The advantage of mobilizing endogenous stem cells over transplantation is tremendous. Cells are generated locally and do not have to migrate extensive distances from the transplant site. Further, there is no possibility of rejection of the mobilized cells, a concern for all transplanted tissue. As far as we are aware of clinical trials on the effect of mobilizing endogenous stem cells have been undertaken. Rodent studies have suggested measurable improvements in learning and memory (eg: [64]) and, given that many of the strategies required to mobilize stem cells are similar to therapy considered useful in stemming the effects of aging, we have no doubt that these experiments will be undertaken.

10.   Cell replacement

A therapeutic approach that is perfectly suited to stem cells comprises the culture of cells prior to their introduction into patients for therapeutic benefit. The ability to grow cells in culture for prolonged periods (eg: [11, 12]), the ability to introduce genes into stem cells (eg: [75]), and the ability of stem cells to migrate long distances in intact and damaged brains (eg: [76]) have led to the hypothesis that stem cells may serve as cellular therapy in a variety of disorders. Some of the most fundamental questions relate to the capacity of existing tissue structures to serve as templates for the incoming stream of new cells. Are fresh young cells the primary deficit in severely damaged or aging organs? Or does a more global breakdown in organization preclude this therapeutic approach? These questions are being addressed and trials with stem cells are in progress [77, 78].

A potential disadvantage of using multipotent cells is the inability to completely regulate the differentiation of cells *in vivo*. Thus multipotent stem cells may differentiate into glial cells when only neuronal replacement is required or vice versa. In these circumstances where replacement of a single cell type is required more restricted precursors may be useful (reviewed in [78]). Alternatively stem cells may be induced to differentiate *in vitro* to produce either specialized cell types for reintroduction or to assemble functional organs for transplantation. A second potential disadvantage of transplanting stem cells is the increasing evidence that stem cells themselves are not a homogenous population, displays some regional specificity and differ in their differentiation ability. Thus multiple lines of stem cells from appropriate regions may be required for appropriate therapy.

Nevertheless transplanting stem cells or differentiated cells derived from passaged stem cells offers some advantages and both functional and behavioral improvements have been demonstrated in rodent models of Parkinson's disease [79] and stroke [80]. Currently clinical trials are underway with passaged cells though no published data are available. In addition to stem cells, more differentiated but not fully differentiated cells (intermediate precursors) can also be used for transplantation. These cells offer the advantage of obtaining more selective populations of cells without the possibility of large numbers of undifferentiated stem cells persisting in the brain. These cells also might respond to cues present in the adult brain better than more undifferentiated cells. Of the differentiated populations several studies have been directed at glial precursors for myelination or remyelination. Data in rodent models is quite encouraging. In particular, transplants of olfactory ensheathing cells appear to promote axonal regeneration far more than any other glial population [81].

It is difficult to envision stem cell transplant therapy as a global replacement during aging. However, one can imagine focal replacement in specific regions of the brain. The loss of cells in Parkinson's disease is relatively localized and the loss of forebrain cholinergic neurons appears to be a common deficit in age related dementias. Perhaps such focal replacement will be useful in promoting successful aging.

11.   Stem cells or precursor cells as drug delivery agents

Delivering drugs across the blood-brain barrier has been difficult and stem cells and precursor cells offer the potential of bypassing this barrier. Two major observations have suggested that this may be a viable option. Astrocytes and glial precursors have been shown to migrate extensively and integrate seamlessly into the host parenchyma (eg: [82]). Furthermore, stem cells have been shown to migrate in a directed manner towards tumors and other lesions providing the potential of localized delivery of antitumor agents [83]. A second breakthrough has been in the development of new vectors that will infect human cells with high efficiency and further maintain expression for prolonged periods (reviewed in [84, 85]). Together, these advances suggest that stem cells may be a viable drug delivery agent. Which drugs, growth factor or peptide will be the agent of choice remains to be determined.

12.   Stem cells for gene and drug discovery

Mobilizing endogenous stem cells or providing exogenous stem cells represent direct therapy. An alternative and as yet relatively unexplored use of stem cells is to utilize the ability to obtain large numbers of identified and characterized cells from the nervous system for gene or drug discovery.

Many methods of whole genome scanning exist. These include SAGE, differential display, gene chip analysis, whole library sequencing, etc. (reviewed in [86]). All of these methods require large amounts of mRNA or cDNA from purified, characterized populations of cells and stem cells and precursor cell populations offer the ability to obtain such numbers of cells. Indeed such experiments are already in progress and profiles of the properties of these cells is rapidly accumulating (eg: [87]). In this context it is intriguing to note that presenilins, identified on the basis of their role in the etiology of Alzheimers, have now been shown to regulate notch proteolytic processing also (see for example [88, 89], and be expressed by stem cells. Neuroepithelial stem cells cultured from the ps1-/- embryos show a decreased self renewal capacity and premature differentiation (Hitoshi et al., 2000 Society of Neurosci. Abst. 23.7). Thus such scans offer the possibility of identifying functionally relevant genes.

Another strategy that has not been fully utilized is that of using well-characterized cell populations for drug discovery. Virtually 90% of drugs used today act to modulate neurotransmitter receptors and having cell lines that express these receptors would be useful. However, human cells lines for the CNS are not readily available. CNS neuronal tumors are rare and glial tumors of the oligodendrocyte lineage are difficult to maintain in culture, undergo a fibroblastic transformation and no longer resemble glial cells. The availability of dividing precursor cells offers the ability to generate such lines. We and others have generated such lines from rodent and human CNS using regulatable oncogenes [23, 90, 91]. The demonstration that telomerase activation can be used to generate human cell lines that remain karyotypically stable offers another possibility for obtaining useful CNS stem cell lines.

A third strategy that can conceivably be used with stem cells is somatic cell genetics to identify functionally important genes. ES cells have been used for promoter trap, enhancer trap and promoter reporter screens because of the ability to maintain cells for prolonged periods in culture. These and other functional screens have identified novel signaling pathways, new genes or unexpected functions of known genes. Several investigators have extended these approaches to cell lines and to non-transformed somatic cells that can be grown for prolonged periods in culture (reviewed in [92]). The ability to obtain human neural lines and the recent demonstration that glial cells are spontaneously immortal and do not undergo senescence should allow us to use existing tools to identify neural genes that regulate growth proliferation and differentiation.

13.   Closing remarks

Normal CNS development involves the sequential differentiation of multipotent stem cells to precursor cells through additional stages of development to generate differentiated cells of the adult nervous system. Stem cells persist in the adult and are capable of contributing to repair and regeneration. Alteration of stem cell numbers, self renewal capacity or proliferative capacity will have major effects on the appropriate development of the nervous system, in the pool of stem cells present in the adult, and in their response to injury or senescence. In this review we discuss current information about human stem cells and precursor cells, the evidence that they exist in the adult, the properties of adult stem cells that may allow them to be used as successful vehicles for replacement therapy and we describe potential therapeutic strategies.

14.   Conclusion

Interventions for preventive geriatrics and successful aging include a low-fat, low-energy diet with a high content of fruits and vegetable, exercises, and hormone replacement. We submit that successful aging of the brain can mediate overall successful aging because the nervous system either directly or indirectly controls energy metabolism, hormonal levels, endocrine and cardiac function, as well as controlling all visceral functions through the autonomic system. Modulating the aging process in the brain may therefore regulate age-dependent degradations that occur in all organ systems. Mobilizing endogenous stem cells, providing exogenous cells to provide cellular replacement or delivery of therapeutic molecules provide novel mechanism of regulating the secondary effects of aging on the brain. Recent advances in our ability to obtain large numbers of stem cells, to isolate and maintain them in culture and to manipulate their properties by genetic or epigenetic means provide hope that therapeutic intervention may be a possibility in the near future.

15.   References

1.   Holloszy, J. O. (2000) Mayo Clin Proc 75 Suppl, S3-8; discussion S8-9.

2.   Brody, H. (1992). Acta Neurol Scand Suppl 137, 40-4.

3.   Singer W. (1982). Eds J.G. Nicolls. Pp 203-226.   Springer-Verlag Press NY.

4.   Buc-Caron, M. H. (1995). Neurobiol Dis 2, 37-47.

5.   Sabate, O., Horellou, P., Vigne, E., Colin, P., Perricaudet, M., Buc-Caron, M. H., and Mallet, J. (1995). Nat Genet 9, 256-60.

6.   Svendsen, C. N., Clarke, D. J., Rosser, A. E., and Dunnett, S. B. (1996). Exp Neurol 137, 376-88.

7.   Chalmers-Redman, R. M., Priestley, T., Kemp, J. A., and Fine, A. (1997). Neuroscience 76, 1121-

8.   Svendsen, C. N., ter Borg, M. G., Armstrong, R. J., Rosser, A. E., Chandran, S., Ostenfeld, T., and Caldwell, M. A. (1998). J Neurosci Methods 85, 141-52.

9.   Johansson, C. B., Momma, S., Clarke, D. L., Risling, M., Lendahl, U., and Frisen, J. (1999). Cell 96, 25-34.

10.   Johansson, C. B., Svensson, M., Wallstedt, L., Janson, A. M., and Frisen, J. (1999). Exp Cell Res 253, 733-6.

11.   Vescovi, A. L., Gritti, A., Galli, R., and Parati, E. A. (1999). J Neurotrauma 16, 689-93.

12.   Vescovi, A. L., Parati, E. A., Gritti, A., Poulin, P., Ferrario, M., Wanke, E., Frolichsthal-Schoeller, P., Cova, L., Arcellana-Panlilio, M., Colombo, A., and Galli, R. (1999). Exp Neurol 156, 71-83.

13. Carpenter, M. K., Cui, X., Hu, Z. Y., Jackson, J., Sherman, S., Seiger, A., and Wahlberg, L. U. (1999). Exp Neurol 158, 265-78.

14. Quinn, S. M., Walters, W. M., Vescovi, A. L., and Whittemore, S. R. (1999). J Neurosci Res 57, 590-602.

15. Piper, D. R., Mujtaba, T., Rao, M. S., and Lucero, M. T. (2000). J Neurophysiol 84, 534-48.

16. Kukekov, V. G., Laywell, E. D., Suslov, O., Davies, K., Scheffler, B., Thomas, L. B., O'Brien, T. F., Kusakabe, M., and Steindler, D. A. (1999). Exp Neurol 156, 333-44.

17. Roy, N. S., Benraiss, A., Wang, S., Fraser, R. A., Goodman, R., Couldwell, W. T., Nedergaard, M., Kawaguchi, A., Okano, H., and Goldman, S. A. (2000). J Neurosci Res 59, 321-31.

18. Hans Lu. (1982). developmental Neurobiology Pages 291. Raven press

19. Lee, J. C., Mayer-Proschel, M., and Rao, M. S. (2000a). Glia 30, 105-21.

20. Rao, M. S. (1999). Anat Rec 257, 137-48.

21. Ling, Z. D., Potter, E. D., Lipton, J. W., and Carvey, P. M. (1998). Exp Neurol 149, 411-23.

22. Wang, S., Roy, N. S., Benraiss, A., and Goldman, S. A. (2000). Dev Neurosci 22, 167-76.

23. Li, R., Thode, S., Zhou, J., Richard, N., Pardinas, J., Rao, M. S., and Sah, D. W. (2000). J Neurosci Res 59, 342-52.

24. Piper D. R.,Mujtaba T., Keyoung H., Roy N. S.,, Goldman S. A., Rao M. S., and Lucero M. T. (2001). J. Neurosci. Submitted.

25. Rao, M. S., Noble, M., and Mayer-Proschel, M. (1998). Proc Natl Acad Sci U S A 95, 3996-4001.

26. Roy, N. S., Wang, S., Harrison-Restelli, C., Benraiss, A., Fraser, R. A., Gravel, M., Braun, P. E., and Goldman, S. A. (1999). J Neurosci 19, 9986-95.

27. Scolding, N. J., Rayner, P. J., and Compston, D. A. (1999). Neuroscience 89, 1-4.

28. Prabhakar, S., D'Souza, S., Antel, J. P., McLaurin, J., Schipper, H. M., and Wang, E. (1995). Brain Res 672, 159-69.

29. Raine, C. S., Scheinberg, L., and Waltz, J. M. (1981). Lab Invest 45, 534-46.

30. Scolding, N. J., Rayner, P. J., Sussman, J., Shaw, C., and Compston, D. A. (1995). Neuroreport 6, 441-5.

31. Mi, H., and Barres, B. A. (1999). J Neurosci 19, 1049-61.

32. Seidman, K. J., Teng, A. L., Rosenkopf, R., Spilotro, P., and Weyhenmeyer, J. A. (1997). Brain Res 753, 18-26.

33. Tang, D. G., Tokumoto, Y. M., Apperly, J. A., Lloyd, A. C., and Raff, M. C. (2001). Science 291, 868-871.

34. Mathon, N. F., Malcolm, D. S., Harrisingh, M. C., Cheng, L., and Lloyd, A. C. (2001). Science 291, 872-875.

35. Laywell, E. D., Rakic, P., Kukekov, V. G., Holland, E. C., and Steindler, D. A. (2000). Proc Natl Acad Sci U S A 97, 13883-8.

36. Kondo, T., and Raff, M. (2000). Science 289, 1754-7.

37. Rao M.S. and Sommer L. (2001). Progress in neurobiology. Invited review

38. Morrison, S. J., Wandycz, A. M., Akashi, K., Globerson, A., and Weissman, I. L. (1996). Nat Med 2, 1011-6.

39. Lemaire, V., Koehl, M., Le Moal, M., and Abrous, D. N. (2000). Proc Natl Acad Sci U S A 97, 11032-7.

40. Martin, K., Potten, C. S., and Kirkwood, T. B. (2000). Ann N Y Acad Sci 908, 315-8.

41. Smith, J., Ladi, E., Mayer-Proschel, M., and Noble, M. (2000). Proc Natl Acad Sci U S A 97, 10032-7.

42. Ostenfeld, T., Caldwell, M. A., Prowse, K. R., Linskens, M. H., Jauniaux, E., and Svendsen, C. N.

(2000). Exp Neurol 164, 215-26.

43. Fu, W., Killen, M., Pandita, T., and Mattson, M.P. (2000). J. Mol. Neurosci. 14, 3-15.

44. Klapper, W., Shin, T., and Mattson, M.P. (2001). J. Neurosci. Res. 64(3) 252-260.

45. Zhen, X., Uryu, K., Cai, G., Johnson, G. P., and Friedman, E. (1999). J Gerontol A Biol Sci Med Sci 54, B539-48.

46. Ravati, A., Ahlemeyer, B., Becker, A., and Krieglstein, J. (2000). Brain Res 866, 23-32.

47. Seckl, J. R. (2000). J Neuroendocrinol 12, 709-10.

48. Belugin, S., Akino, K., Takamura, N., Mine, M., Romanovsky, D., Fedoseev, V., Kubarko, A., Kosaka, M., and Yamashita, S. (1999). Thyroid 9, 837-43.

49. Ciani, E., Groneng, L., Voltattorni, M., Rolseth, V., Contestabile, A., and Paulsen, R. E. (1996). Brain Res 728, 1-6.

50. Lee, J., Duan, W., Long, J.M., Ingram, D.K., and Mattson, M.P. (2000b). J. Mol. Neurosci. 15, 99-108.

51. Chiasson, B. J., Tropepe, V., Morshead, C. M., and van der Kooy, D. (1999). J Neurosci 19, 4462-71.

52. Marmur, R., Mabie, P. C., Gokhan, S., Song, Q., Kessler, J. A., and Mehler, M. F. (1998). Dev Biol 204, 577-91.

53. Kalyani, A. J., Mujtaba, T., and Rao, M. S. (1999). J Neurobiol 38, 207-24.

54. Uchida, N., Buck, D. W., He, D., Reitsma, M. J., Masek, M., Phan, T. V., Tsukamoto, A. S., Gage, F. H., and Weissman, I. L. (2000). Proc Natl Acad Sci U S A 97, 14720-5.

55. Hastings, N. B., and Gould, E. (1999). J Comp Neurol 413, 146-54.

56. Cameron, H. A., and Gould, E. (1996). J Comp Neurol 369, 56-63.

57. Parent, J. M., Yu, T. W., Leibowitz, R. T., Geschwind, D. H., Sloviter, R. S., and Lowenstein, D. H. (1997). J Neurosci 17, 3727-38.

58. Cameron, H. A., and McKay, R. D. (1999). Nat Neurosci 2, 894-7.

59. Gould, E., and Cameron, H. A. (1996). Dev Neurosci 18, 22-35.

60. Kempermann, G., Kuhn, H. G., and Gage, F. H. (1997). Proc Natl Acad Sci U S A 94, 10409-14.

61. Dash, P. K., Mach, S. A., and Moore, A. N. (2001). J Neurosci Res 63, 313-319.

62. Gould, E., Tanapat, P., Rydel, T., and Hastings, N. (2000). Biol Psychiatry 48, 715-20.

63. Madeira, M. D., Paula-Barbosa, M., Cadete-Leite, A., and Tavares, M. A. (1988). J Hirnforsch 29, 643-50.

64. Derrick, B. E., York, A. D., and Martinez, J. L. (2000). Brain Res 857, 300-7.

65. Corotto, F. S., Henegar, J. R., and Maruniak, J. A. (1994). Neuroscience 61, 739-44.

66. Smith, M. T., Pencea, V., Wang, Z., Luskin, M. B., and Insel, T. R. (2001). Horm Behav 39, 11-21.

67. Kuhn, H. G., Winkler, J., Kempermann, G., Thal, L. J., and Gage, F. H. (1997). J Neurosci 17, 5820-9.

68. Martens DJ, Tropepe V, van Der Kooy D. (2000). J Neurosci. 20(3), 1085-95.

69. Magavi, S. S., Leavitt, B. R., and Macklis, J. D. (2000). Nature 405, 951-5.

70. Shihabuddin, L. S., Horner, P. J., Ray, J., and Gage, F. H. (2000). J Neurosci 20, 8727-35.

71. Panicker M. and Rao M. S. (2000). In " Stem cells". Eds. D. Marshiak and D. Gotlieb. CSH Press.

72. Martin, G. M. (1979). Mech Ageing Dev 9, 385-91.

73. Martin, G. M. (2000). Ann N Y Acad Sci 908, 1-13.

74. Martin, G. M. (2000). Exp Gerontol 35, 439-43.

75. Liu, Y., Himes, B. T., Solowska, J., Moul, J., Chow, S. Y., Park, K. I., Tessler, A., Murray, M., Snyder, E. Y., and Fischer, I. (1999). Exp Neurol 158, 9-26.

76. Flax, J. D., Aurora, S., Yang, C., Simonin, C., Wills, A. M., Billinghurst, L. L., Jendoubi, M., Sidman, R. L., Wolfe, J. H., Kim, S. U., and Snyder, E. Y. (1998). Nat Biotechnol 16, 1033-9.

77. Bjorklund, A., and Lindvall, O. (2000). Nat Neurosci 3, 537-44.

78. Rao, M. S., and Mayer-Proschel, M. (2000) Prog Brain Res 128, 273-92.

79.   Svendsen, C. N., Caldwell, M. A., Shen, J., ter Borg, M. G., Rosser, A. E., Tyers, P., Karmiol, S., and Dunnett, S. B. (1997). Exp Neurol 148, 135-46.

80.   Kondziolka, D., Wechsler, L., Goldstein, S., Meltzer, C., Thulborn, K. R., Gebel, J., Jannetta, P., DeCesare, S., Elder, E. M., McGrogan, M., Reitman, M. A., and Bynum, L. (2000). Neurology 55, 565-9.

81.   Ramon-Cueto, A., Cordero, M. I., Santos-Benito, F. F., and Avila, J. (2000). Neuron 25, 425-35.

82.   Keirstead, H. S., Ben-Hur, T., Rogister, B., O'Leary, M. T., Dubois-Dalcq, M., and Blakemore, W. F. (1999). J Neurosci 19, 7529-36.

83.   Yandava, B. D., Billinghurst, L. L., and Snyder, E. Y. (1999). Proc Natl Acad Sci U S A 96, 7029-34.

84.   Wu, N., and Ataai, M. M. (2000). Curr Opin Biotechnol 11, 205-8.

85.   Hu, W. S., and Pathak, V. K. (2000). Pharmacol Rev 52, 493-511.

86.   Rao M.S. and Kamb A. (2000). In "Stem cells and CNS Development". Eds Rao MS. Humana Press. NY.

87.   Kelly, D. L., and Rizzino, A. (2000). Mol Reprod Dev 56, 113-23.

88.   Berezovska, O., Jack, C., McLean, P., Aster, J. C., Hicks, C., Xia, W., Wolfe, M. S., Weinmaster, G., Selkoe, D. J., and Hyman, B. T. (2000). Ann N Y Acad Sci 920, 223-6.

89.   Zhang Z, Nadeau P, Song W, Donoviel D, Yuan M, Bernstein A, Yankner BA. (2000). Nat Cell Biol. 2(7), 463-5.   No abstract available.

90.   Sah, D. W., Ray, J., and Gage, F. H. (1997). Nat Biotechnol 15, 574-80.

91.   Raymon, H. K., Thode, S., Zhou, J., Friedman, G. C., Pardinas, J. R., Barrere, C., Johnson, R. M., and Sah, D. W. (1999). J Neurosci 19, 5420-8.

92.   Sedivy, J. M., and Dutriaux, A. (1999). Trends Genet 15, 88-90.

CHAPTER 6

STEM CELLS AND NEURODEGENERATIVE DISORDERS

MARK P. MATTSON, NORMAN J. HAUGHEY,
AIWU CHENG and MAHENDRA S. RAO

Table of contents

1.   Aging of the nervous system: cellular and molecular correlates

Before considering the impact of aging on neural progenitor cells (NPC), it is important to have a basic understanding of changes that occur in the nervous system during aging and in age-related neurological disorders. As is the case in other organ systems, aging in the nervous system is associated with increased levels of oxidative damage to proteins and DNA; accumulation of protein and lipid by-products (e.g., lipofuscin and advanced glycation end products), and impaired cellular energy metabolism [1, 2]. Alterations in signal transduction pathways involving neurotransmitters, trophic factors and cytokines that are involved in regulating neuronal excitability and plasticity are also subject to modification by aging [3]. The cellular and molecular changes that occur in the nervous system during "usual" aging may predispose neurons to degeneration in disorders such as Alzheimer's disease (AD) [4], Parkinson's disease (PD) [5], Huntington's disease (HD) [6] and amyotrophic lateral sclerosis (ALS) [7]. Such changes may also render neurons vulnerable to acute insults such as ischemic stroke and traumatic injury.

Age-related structural changes that have been documented in aging nervous systems include nerve cell death, dendritic retraction and expansion, synapse loss and remodeling, and glial cell reactivity [8, 9]. Decreases in synapse numbers in some regions of the nervous system may occur during aging, but may be offset by increases in synaptic size.

Stem Cells: A Cellular Fountain of Youth.   Ed. by Mark P. Mattson and Gary Van Zant. 117 — 139

These changes are often associated with alterations in cytoskeletal proteins and the deposition of insoluble proteins such as amyloid in the extracellular space. Cytoskeletal changes include: increased levels of phosphorylation of the microtubule-associated protein tau which occur in some brain regions, particularly those involved in learning and memory (e.g., hippocampus and basal forebrain); calcium-mediated proteolysis of proteins such as MAP-2 and spectrin; oxidative modification of various cytoskeletal proteins including glycation and covalent binding of the lipid peroxidation product 4-hydroxynonenal; and an increase in levels of glial fibrillary acidic protein [10-13]. As in other organ systems, vessels that supply blood to the brain and spinal cord are vulnerable to age-related atherosclerosis and arteriosclerosis. Such changes render the blood vessels susceptible to occlusion or rupture, which can manifest as stroke, a major cause of disability and death in the elderly population [14]. Age-dependent cerebral vascular changes are strongly linked to heart disease, hypertension, hyperlipidemia, diabetes mellitus, and smoking.

Several different alterations that occur during usual aging of the nervous system are greatly enhanced in neurodegenerative disorders. For example, levels of lipid peroxidation products such as lipid peroxides and the aldehyde 4-hydroxnonenal are significantly elevated in vulnerable brain regions of AD and PD patients, and in spinal cords of ALS patients [15]. Levels of protein oxidation, measured as protein carbonyls, are significantly elevated in the brains of aged rodents, and such age-associated protein oxidation can be prevented by administration of antioxidants [16]. Protein oxidation is also increased in brain tissues from AD and PD patients; oxidation of proteins impairs their function and is therefore likely to contribute to age-related cellular dysfunction and neuronal degeneration. Two proteins that have been shown to be heavily glycated in AD are amyloid β-peptide (Aβ) and tau, the major components of plaques and neurofibrillary tangles, respectively [10].

Mitochondrial DNA damage occurs during aging and this may be accelerated in neurodegenerative disorders; mitochondrial DNA is particularly vulnerable because high levels of free radicals are generated locally, and because mitochondria do not possess effective DNA repair systems [17]. Although very little unrepaired nuclear DNA damage occurs in the nervous system during usual aging, extensive DNA damage occurs in vulnerable neurons in AD, PD and HD patients [18, 19]. This DNA damage may be caused by reactive oxygen molecules such as hydroxyl radical and peroxynitrite, and by impaired DNA repair resulting from genetic or dietary factors [20].

An important feature of neurons that may contribute to their vulnerability during aging is their excitability, which is regulated by a complex array of neurotransmitters and ion channels [2]. Neurons express voltage-dependent sodium and calcium channels, ligand-gated ion channels, and various membrane ion transporters. Numerous age-related alterations in electrophysiological parameters have been described in neurons in rodents and humans. The alterations range from increased thresholds for induction of action potentials in cranial nerves, to increased afterhyperpolarizations and impaired long-term potentiation of synaptic transmission in hippocampal neurons [21]. Particularly important are alterations in neuronal systems that regulate calcium homeostasis, which may result in increased levels of intracellular calcium leading to neuronal dysfunction and degeneration [22]. Overactivation of glutamate receptors is thought to contribute to age- and disease-related disruption of cellular calcium homeostasis in a process called "excitotoxicity" [2].

An array of alterations in neuroendocrine and neuroimmune systems occur during aging. Changes in levels of steroid horomones, and regulation of their production by the brain – hypothalamus – pituitary system, have been documented in aging animals and humans [23, 24], The changes include: alterations in the diurnal regulation of circulating glucocorticoids with an increase in mean levels; a marked decrease in estrogen levels that occurs in women after menopause; and a progressive decrease in testosterone levels in men beyond the age of 20. Epidemiological studies have shown that there is a reduced risk of AD in postmenopausal women who take estrogen replacement therapy [25], consistent with a role for estrogen in warding off this disease. The brain is intimately linked with the immune system, and reciprical interactions between nervous and immune systems are likely to contribute to age-related neurological and immune disorders [26]. For example, a decline in peripheral immune function during aging may lead to autoimmune-like phenomena in the brain resulting in activation of microglia [27]. Inflammatory responses, involving complement activation and microglial cytokine production, are thought to facilitate the neurodegenerative process in AD [28].

Neurons and glial cells produce proteins that serve the function of promoting the survival and growth of neurons, and such neurotrophic factors may protect the neurons against injury- and age-related degeneration. Indeed, neurotrophic factors such as nerve growth factor (NGF), basic fibroblast growth factor (bFGF), brain-dervied neurotrophic factor (BDNF), and insulin-like growth factor (IGF) can protect neurons in cell culture and *in vivo* against a variety of insults relevant to the pathogenesis of AD, stroke and PD [3, 29]. Three general mechanisms whereby neurotrophic factors prevent neuronal degeneration are by enhancing antioxidant systems, inducing expression of anti-apoptotic proteins, and stabilizing cellular calcium homeostasis.

It is quite remarkable that many persons can live into their 80s, 90s and 100s without major neurological dysfunction. An intriguing feature of neurotrophic factors is that their expression is increased by activity in neuronal circuits. Rats and mice reared in "intellectually enriched" environments exhibit expansion of dendritic arbors and increased numbers of synapses in hippocampus and certain regions of cerebral cortex, and improved learning and memory ability [30]. Epidemiological data suggest that humans with "active" minds have a reduced risk for developing AD as they age [31]. Recent findings have also shown that diet can modulate brain aging and risk of neurodegenerative disorders [32]. Thus, dietary restriction (reduced calorie intake) can increase resistance of neurons in the brain to dysfunction and death in experimental models of AD, PD, HD and stroke [33-35]. The mechanism underlying the beneficial effects of dietary restriction involves stimulation of the expression of "stress proteins" and neurotrophic factors. When taken together with data showing that neurotrophic factors can protect brain neurons against insults relevant to usual brain aging and neurodegenerative disorders, these findings suggest a homeopathic mechanism of successful brain aging in which brain activity and dietary restriction stimulate the expression of neurotrophic factors which, in turn, promote neuronal growth and plasticity (Figure 1).

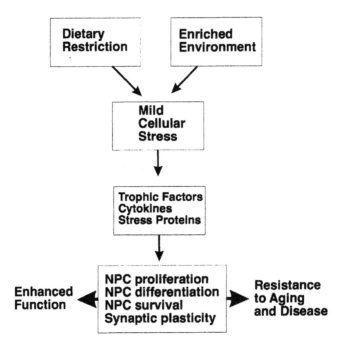

Figure 1.   Working model of the cellular and molecular mechanisms whereby dietary restriction and environmental enrichment enhance neuronal plasticity. Dietary restriction induces a mild energetic stress on neural cells, while environmental enrichment also stresses cells as a result of increased activity. Cells respond to the mild stress by upregulating the expression of genes that encode proteins that promote cell survival and plasticity; examples include antioxidant enzymes, heat-shock proteins, anti-apoptotic proteins and neurotrophic factors. Neural progenitor cells may respond to dietary restriction and environmental enrichment by increasing their survival and differentiation. Mature neurons may respond to the mild stress by increasing their resistance to injury and enhancing their synaptic connections. In these ways, dietary restriction and environmental enrichment increase resistance of neurons to aging and age-related disease.

2.   Pathogenesis of age-related neurodegenerative disorders

As the average life expectancy increases, so too does the incidence of neurodegenerative disorders such as AD, PD and stroke (Figure 2). In fact, the latter three disorders place a greater financial and social burden on our society than do either coronary artery disease or cancer. Considerable progress has been made in identifying genetic and environmental factors that cause, or increase risk of, these and other neurodegenerative disorders. A general picture is emerging in which age-related increases in cellular oxidative stress and impairment of energy metabolism, combine with one or more disease-initiating factors to cause intracellular neurodegenerative cascades in neurons. The neurodegenerative cascades often involve one or more of the following alterations:

increased oxidative damage to DNA, proteins and lipids; perturbed cellular ion (particularly calcium) homeostasis; increased production and/or activation of apoptosis effector proteins such as caspases, Par-4 and pro-apoptotic members of the Bcl-2 family of proteins [2, 36]. The reasons that specific populations of neurons are vulnerable in a particular disorder (for example, hippocampal neurons in AD *versus* substantia nigra neurons in PD) are beginning to be understood and may be related to specific genetic and environmental factors. Before considering the possible involvement of alterations in neural stem cells in the pathogenesis of neurodegenerative disorders, and the potential application of stem cell therapies to these disorders, we briefly describe several of the most prominent neurodegenerative conditions.

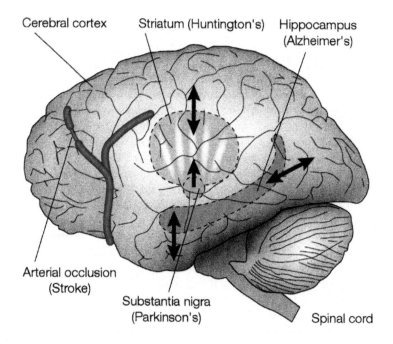

Figure 2.   Examples of neurodegenerative brain disorders and the brain regions involved. Alzheimer's disease involves degeneration of neurons in the hippocampus and associated regions of cerebral cortex that are involved in learning and memory. Motor dysfunction results from degeneration of neurons in the substantia nigra in Parkinson's disease and in the striatum (caudate and putamen) in Huntington's disease. Occlusion or rupture of a blood vessel results in an ischemic stroke, which damages neurons supplied by that vessel.

AD involves progressive impairment of cognition and emotional disturbances that result from degeneration of synapses and death of neurons in limbic structures such as hippocampus and amygdala, and associated regions of cerebral cortex. A prominent abnormalilty in AD is accumulation of amyloid plaques formed by aggregates of Aβ, a 40-42 amino acid fragment generated by proteolytic processing of the amyloid precursor protein (APP) [2]. DNA damage, caspase activation, and alterations in expression of apoptosis-related genes such as Bcl-2 family members, Par-4 and DNA damage response genes have been documented in neurons associated with amyloid deposits in the brains of AD patients [37, 38]. Studies of cell culture and animal models of AD have shown that Aβ can induce apoptosis of neurons directly and can greatly increase their vulnerability to death induced by conditions such as oxidative stress and reduced energy availability, that occur during aging. Aβ may promote cell death by inducing oxidative stress, mitochondrial dysfunction and DNA damage [4]. A small percentage of cases of AD are caused by mutations in single genes that are inherited in an autosomal dominant manner; persons with such familial forms of AD become symptomatic at a remarkably early age (30s, 40s or 50s) [39]. Mutations in three different genes, can cause early-onset familial AD; one gene encodes APP, a second encodes presenilin-1, and a third encodes presenilin-2 [40]. Mutations in each gene cause an increase in the production of Aβ and its accumulation in the brain. When mutant presenilin-1 protein is expressed in cultured cells and in transgenic and knockin mice, neurons become susceptible to death induced by a variety of insults including trophic factor withdrawal and exposure to Aβ, glutamate and energy deprivation [41-43]. Mutant presenilin-1 causes a disturbance in cellular calcium homeostasis which appears to be a pivotal event in promoting DNA damage and degeneration of neurons [36].

Progressive deterioration of neurons that control muscle movements is responsible for several prominent disorders of aging including PD, HD and ALS. PD patients have profound motor dysfunction as the result of degeneration of dopaminergic neurons in brain region called the substantia nigra. There is evidence that the dopaminergic neurons undergo apoptosis and that DNA damage is involved in the cell death process [5]. A prominent alteration in PD is a deficit in mitochondrial complex I. Environmental and genetic factors may sensitize dopaminergic neurons to age-related increases in oxidative stress and energy deficits; environmental toxins including pesticides such as rotenone and other chemicals are implicated [44]. A very small percentage of cases of PD are inherited in a dominant manner and may result from mutations in genes encoding either α-synuclein or parkin [45]. α-synuclein is a component of brain lesions called Lewy bodies, and may promote dysfunction and death of dopaminergic neurons by impairing synaptic transmission and inducing apoptosis [46, 47].

HD is a purely inherited disease that involves progressive degeneration of neurons in the striatum (caudate nucleus and putamen) resulting in uncontrollable body movements. It is caused by expansions of a trinucleotide (CAG) sequence in the huntingtin gene producing a protein containing increased polyglutamine repeats [6]. The mutant huntingtin protein may promote selective degeneration of striatal neurons by facilitating apoptosis as suggested by alterations in caspase activities in patients and in animal models of HD [48]. Moreover, inhibition of caspase-1 was reported to slow disease progression in a transgenic mouse model of HD [49]. HD is one of several disorders caused by trinucleotide expansions in genes. Other examples include Kennedy's disease which caused by polyglutamine expansions in the testosterone receptor [50] and spinocerebellar ataxia type 1 which is caused by trinucleotide expansions in the gene encoding ataxin 1 [51]. Although the affected genes are widely expressed in tissues throughout the body, the trinucleotide expansions have adverse effects on a limited number of cell types, with postmitotic cells such as neurons and muscle cells being most vulnerable.

Amyotrophic lateral sclerosis (ALS) results in progressive paralysis due to degeneration of motor neurons in the spinal cord. The neurodegenerative process involves increased oxidative stress, overactivation of glutamate receptors, cellular calcium overload and DNA damage [52]. Most cases of ALS are sporadic, but some cases are caused by mutations in the antioxidant enzyme Cu/Zn-superoxide dismutase; the mutations do not decrease antioxidant activity of the enzyme, but instead result in gain of an adverse proapoptotic activity which may involve increased peroxidase activity. The DNA damage in ALS is associated with increased mitochondrial localization of Bax and decreased association of Bcl-2 [7]. The ability of overexpression of Bcl-2 and administration of caspase inhibitors to delay motor neuron degeneration and death in Cu/Zn-SOD mutant mice suggests a major role for apoptosis in this disease [53, 54]. Studies of spinal cord lower motor neurons and cortical neurons from Cu/Zn-SOD mutant mice suggest that the mutations may promote neuronal degeneration by enhancing oxyradical production and disrupting cellular calcium homeostasis [55].

Neuronal degeneration resulting from acute brain insults such as ischemic stroke and traumatic injury is responsible for much death and disability worldwide. Neurons in the brain region surrounding the central core of an ischemic stroke exhibit DNA damage and activation of the DNA damage-responsive proteins PARP and Ku80 [56]. In rodent stroke models, neurons in the ischemic penumbra exhibit morphological and molecular changes consistent with apoptosis including caspase activation, expression of pro-apoptotic genes and release of cytochrome c. Stabilization of mitochondrial function and inhibition or knockout of caspases results in reduced brain damage after a stroke [14]. Analyses of brains and spinal cords of patients that died after traumatic injuries have documented apoptosis-related changes in neurons including the presence of DNA strand breaks, caspase activation and increased Bax and p53 expression [57, 58]. Intraventricular administration of a caspase-3 inhibitor prior to injury reduces cell death and improves symptoms, and mice expressing a dominant negative inhibitor of caspase-1 exhibit reduced brain damage and free radical production after traumatic injury [59], suggesting a central role for caspases in a brain injury model,

3.   Neural stem cell biology

Detailed information on the cellular and molecular biology of various types of stem cells can be found within the other chapters in this volume, and will therefore only be touched upon briefly here. During the very early (blastocyst) period of development, all cells of the embryo are capable of generating an entire organism; these so-called embryonic stem (ES) cells are thus defined by their potential to generate the entire organism. ES cells isolated from the blastula can be cultured indefinitely *in vitro*; they are characterized by the expression of specific markers, high levels of telomerase activity, and responsiveness to specific growth factors and cytokines [60-63].

Restricted precursor cells capable of acquiring neuronal and glial phenotypes can be isolated from mouse ES cells [64]. A neural stem cell or neural progenitor cell (NPC) can be defined as a cell capable of self-renewal that can undergo asymmetric division to generate differentiated neurons or glia. Progenitor/blast cells therefore have a limited self-renewing capacity and a relatively restricted repertoire of differentiation [65, 66]. Using these definitions, multipotent stem cells have been isolated from most tissues, and several stages of stem cell to progenitor cell differentiation have been identified. It is important to emphasize that NPC are not simply a feature of embryonic development but are also present in adult brain and spinal cord where they may play important roles in the normal homeostasis of the nervous system. For example, in the adult brain populations of NPC located in the subventricular zone form olfactory bulb cells and NPC in the dentate gyrus of the hippocampus can give rise to neurons or astrocytes.

Recent findings suggest that adult stem cells are involved in normal tissue homeostasis, and in aging and disease. Studies in which mitotic cells are labeled with bromodeoxyuridine and other DNA precursors have not only been instrumental in establishing the presence of stem cells in various tissues throughout the body, they have also provided insight into their localization, the phenotypes of their differentiated progeny, and their turnover rates. Such studies have revealed a tremendous heterogeneity in the kinetics of stem cell production, differentiation and survival among tissues. For example, hematopoietic stem cells have a very high proliferation rate and most of their progeny have a relatively short lifespan on the order of days. In contrast, NPC in the brain have a slower proliferation rate and a longer lifespan (weeks to years). In order to study NPC in the context of aging, it is important to understand the molecular mechanisms that regulate NPC behaviors.

4.   Regulation of NPC proliferation, differentiation and survival

During development of the nervous sytem, the numbers of NPC and the process of NPC differentiation are carefully regulated to meet the demands of developing neuronal circuits, and it is therefore not surprising that complex feedback cues have developed to maintain appropriate pools of undifferentiated NPC and differentiated progeny. A NPC may remain quiescent and simply not enter the cell cycle, which is important for maintaining a reserve pool of cells for use in times of stress or at later stages of development. A NPC may undergo apoptosis and not contribute to further development, which may be the norm in the adult brain where turnover of differentiated cells (neurons and glial cells) is very low. NPC may undergo symmetric cell divisions to self renew or undergo terminal differentiation, or they may undergo asymetric cell divisions to generate differentiated progeny as well as maintain a pool of stem cells. A dynamic balance between proliferation, survival and differentiation signals ensures that an appropriate balance between NPC and differentiated cells is maintained throughout development and adult life. NPC from the embryonic or adult nervous system can be isolated and maintained in culture as "neurospheres" (Figure 3). When plated on an adhesive substrate, the self-renewing NPC remain within the central sphere, while newly differentiated neurons and glia migrate radially across the growth substrate. Such neurosphere preparations have proven valuable in elucidating cellular and molecular mechanisms that regulate NPC fate.

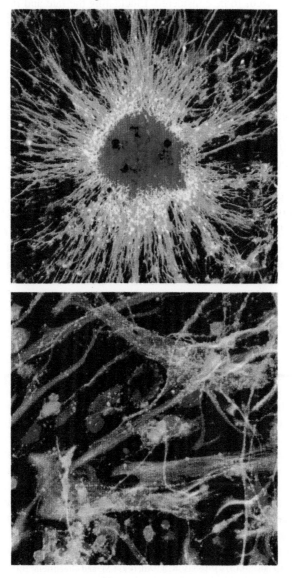

Figure 3.   Neurospheres represent a dynamic cellular system for studies of the molecular and cellular regulation of NPC proliferation, differentiation and survival. The upper micrograph shows a confocal laser scanning micrograph of a neurosphere from embryonic human cerebral cortex at low magnification. The neurosphere was triple-labeled with markers of cell nuclei (red), astrocytes (white) and neurons (green). Note that the center of the sphere contains a very high density of cells that are not labeled by markers of differentiated astrocytes and neurons, and that the differentiated neurons and astrocytes have migrated radially away from the center of the neurosphere. The lower micrograph is a higher magnification view of cells that have migrated away from the center of the sphere.

In the adult brain, populations of NPC located in the subventricular zone and the dentate gyrus of the hippocampus turn over continuously such that most, if not all, newly generated cells die within a certain time period in the postmitotic state. Most subventricular NPC form olfactory bulb cells which subserve their function and then are replaced by newly-generated cells, although some of this population of NPC may also contribute new cells to the cerebral cortex. NPC appear to play particularly important roles in responses of the nervous system to environmental demands or tissue injury. For example, the proliferation and/or survival of NPC is increased in response to environmental enrichment (intellectual activity) [67, 68], physical activity [69], epileptic seizures [70], traumatic brain injury [71] and ischemia [72]. While multiple signaling pathways can influence stem cell fate under basal conditions and in response to environmental demands, there appear to be a few mechanisms that are widely employed across a range of tissues and in an array of organisms. Conserved intrinsic regulators include telomerase and Bcl-2 family members, while extrinsic regulators include the Notch signaling pathway, growth factors/cytokines and cell adhesion molecules. In simple terms, such signals control two major NPC "decisions", namely, to proliferate or differentiate and to live or die. Thus, stem cells must often "guide" themselves down roads that are bordered by a cancerous state on one side and mortality on the other. For many NPC this road may become increasingly difficult to navigate during aging and in age-related neurodegenerative disorders.

5.   Molecular control of NPC proliferation, differentiation and survival

Signals that control the fate of NPC are complex and, in general, include mechanisms that regulate stem cell populations in other tissues. We will consider only a few of the many important signaling mechanisms, with a focus on those which have been (directly or indirectly) implicated in the pathogenesis of age-related neurodegenerative disorders. One intriguing signaling pathway that controls NPC fate involves a cell surface receptor called Notch (Figure 4). Notch was discovered in *Drosophila* where it is essential for the proper specification of many different cell fates during the processes of oogenesis, myogenesis, neurogenesis and wing and eye development [73]. Four mammalian Notch homologues have been identified (Notch 1-4), with Notch1 playing a prominent role in the determination of NPC fate during development. Gene targeting-based knockout of Notch1 in mice results in embryonic lethality with major abnormalities in somite formation, neurogenesis and hematopoiesis [74]. Notch is an integral membrane protein with a single membrane-spanning domain. an extracellular (N-terminal) domain that contains a series of tandem epidermal growth factor-like repeats and three Lin/Notch repeats (LNR) that function in ligand binding and Notch activation, and an intracellular (C-terminal) region that contains three ankyrin repeats which mediate interactions with other (cytoplasmic) proteins, a "PEST" sequence that regulates protein turnover and a nuclear localization sequence. The major ligand for Notch1 in mammals is called Delta; binding of Delta to Notch1 results in proteolytic cleavage of Notch at a site located at the cytoplasmic face of the membrane, resulting in release of a Notch C-terminal fragment. The notch C-terminal fragment has been shown to translocate to the nucleus where it is thought to modulate transcription. In addition, extranuclear mechanisms of Notch action have also been proposed based upon interactions with protein such as NF-κB subunits [75].

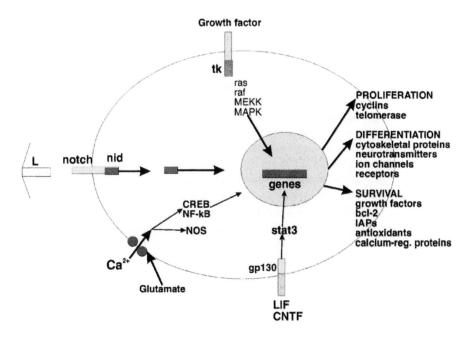

Figure 4. Examples of several prominent signaling mechanisms that can regulate the proliferation, differentiation and/or survival of neural progenitor cells. Notch is a cell surface receptor activated by ligands (L) such as delta resulting in release of a C-terminal notch intracellular domain (nid), which then translocates to the nucleus and induces transcription of genes that promote cell proliferation and survival. Several different growth factors (e.g., bFGF, EGF and BDNF) that activate receptor tyrosine kinases (tk) can affect proliferation, differentiation and/or survival of NPC by activating kinase cascades that ultimately modulate gene expression. Cytokines such as leukemia inhibitor factor (LIF) that activate receptors comprised of gp130 activate a signaling pathway involving stat3 which modulates gene expression in a manner that keeps the cells in a self-renewing state. The fate of NPC can also be regulated by neurotransmitters such as glutamate which induce calcium influx. Calcium activates factors such as CREB and NF-κB that affect gene expression, as well as activating nitric oxide synthase (NOS) resulting in production of nitric oxide. Examples of genes regulated by one or more of these signaling pathways are listed at the right.

Notch can mediate both homotypic and heterotypic cell-cell interactions with both the level and subcellular localization of Notch appearing to play a role in determining cell fate. Among a group of equipotent NPC, those with higher levels of Notch will continue to divide, whereas those with lower levels of Notch will differentiate (Figure 4). For example, during development of the nervous system in *Drosophila* cells expressing low levels of Notch become neurons, while cells with higher levels of Notch become epidermal cells [76]. However, the outcome of Notch signaling can be influenced by growth factors, cell adhesion molecules and other signaling pathways.

Several cytoplasmic proteins have been shown to modify Notch signaling and one such protein that appears to be particularly important in mammalian stem cells is called Numb. By interacting with a cytoplasmic domain of Notch, Numb plays a role in cell fate determination that is antagonistic to that of Notch [77]. During development of the nervous system, Numb plays a pivotal role in controlling asymmetric division of neural progenitor cells by segregating to the daughter cell that remains a progenitor cell [78]. Mice lacking functional Numb die during early development at embryonic day 11.5 and exhibit profound defects in cranial neural tube closure and premature neuron production [79]. Two protein-protein interaction domains have been shown to influence Numb function; a phosphotyrosine-binding (PTB) domain and a proline-rich region (PRR) that functions as a SH3-binding domain [80}. Four different isoforms of Numb that differ in their PTB (lacking or containing an 11 amino acid insert) and PRR (lacking or containing a 48 amino acid insert) domains have been identified. When expressed in NPC, Numb isoforms containing a long PRR promote proliferation whereas isoforms containing a truncated PRR promote cell differentiation [80]. When taken together with emerging evidence that Notch signaling can promote cell survival, the phenotype of the Numb-deficient mice suggests a possible role for Numb in modulating cell survival.

An array of growth factors and cytokines have been identified that can promote the proliferation, differentiation and/or survival of one or more populations of stem cells. For example, epidermal growth factor and basic fibroblast growth factor can maintain brain NPC in a proliferative state [81], while brain-derived neurotrophic factor can facilitate the survival and differentiation of NPC progeny [82]. Insulin-like growth factor-1 is a survival and differentiation factor for several types of progenitor cells including precursors of neurons [83]. Cell adhesion molecules are another class of signaling proteins that play important roles in regulating stem cell behaviors. One very important adhesion signaling system involves integrin receptors which transduce cell-cell and cell-extracellular matrix interactions into various cellular responses including changes in cell proliferation, differentiation and survival. Integrins are activated by binding to extracellular matrix proteins such as laminin or to integrins on the surface of other cells, resulting in an intracellular signaling pathway involving PI3 kinase and Akt kinase. Integrin signaling promotes survival of many different cell types including neurons [84]. NPC express several different integrins which appear to be differentially involved in the regulation of proliferation and cell migration/differentiation. For example, proliferation requires activation of $\alpha5\beta1$ integrins whereas migration requires activation of $\alpha6\beta1$ integrins [85].

The developmental potential of cortical subventricular NPC is differentially affected by EGF and bFGF, with EGF promoting glial cell production and bFGF promoting neuronal differentiation [86, 87]. In addition, BDNF may promote neural differentiation from NPC in the dentate gyrus and cerebral cortex [88]. Novel growth factors and cytokines that affect NPC are also being identified and characterized. For example, secreted forms of amyloid precursor protein (sAPP) can stimulate the proliferation of NPC in culture [89], and can promote the survival of differentiated hippocampal neurons and protect them against excitotoxic and oxidative insults relevant to the pathogenesis of AD and other age-related neurodegenerative disorders [90, 91]. The effects of growth factors on NPC are likely modulated by other factors. For example, proliferation of NPC in response to bFGF requires cystatin C, a protease inhibitor [92]. Interestingly, neurotransmitters can also modify NPC behaviors. Thus, acetylcholine induces proliferation of cortical precursors by a mechanism involving activation of MAP kinase [93]. Glutamate, the major excitatory neurotransmitter in the brain, has also been shown to modulate NPC [94]. The ability of neurotransmitters to affect NPC is consistent with studies showing that neurotransmitters can regulate neurite outgrowth, synaptogenesis and cell survival in the developing brain [95].

A final example of molecular control of stem cell fate concerns an intrinsic mechanism for maintaining genomic stability and responding to DNA damage. Telomeres, the ends of chromosomes, consist of repeats of six-base DNA sequence (TTAGGG) that preserve chromosome integrity and prevent end-to-end fusions. A reverse transcriptase called telomerase is responsible for adding the TTAGGG sequence to the chromosome ends [96]. Telomerase consists of a catalytic subunit (TERT) and an RNA template (TR). Telomerase is expressed in highly proliferative cells throughout the developing embryo and is then dramatically down-regulated as cells differentiate, and is not detectable in many somatic cells in the adult. However, stem cells retain telomerase activity. Studies of cancer cells and somatic cells overexpressing TERT have shown that telomerase can confer upon cells an immortal phenotype. Recent studies have provided evidence that telomerase can promote cell proliferation by protecting telomeres and thereby preventing cell cycle arrest. Moreover, telomerase can prevent apoptosis (programmed cell death) by suppressing DNA damage-induced death cascades involving the tumor suppressor protein p53 [97, 98]. Studies of brain development in mice have correlated a decrease in telomerase activity with decreased neuroblast proliferation, and a decrease in TERT expression with cell differentiation and natural cell death [98, 99]. DNA damage is a trigger for cell cycle arrest and apoptosis, and telomerase is one important mechanism for suppressing such DNA damage. Additional DNA damage response and repair mechanisms are likely to play important roles in the regulation of NPC proliferation and survival including poly (ADP-ribose) polymerase, TRF2, Ku80 and tankyrase [100, 101].

6.   Age-related alterations in NPC

What are the changes that occur in NPC populations during aging? Might changes in NPC influence the rate of aging of the nervous system? How might NPC be used to replace dysfunctional or dead cells and thereby restore function to failing regions of the brain and spinal cord? Remarkably little information is available concerning molecular and cellular changes that occur in NPC during aging. This paucity of information is partly the result of the fact that stem cells comprise only a very small percentage of all cells within a tissue, and it is therefore very difficult to obtain pure populations of stem cells in quantities sufficient to perform many different biochemical or even molecular analyses. Moreover, in many cases the stem cells can not be unambiguously identified *in situ*, making it difficult to perform immunohistochemical or *in situ* hybridization analyses, for example. Nevertheless, the "immortal" nature of stem cells makes their study an imperitive in the field of aging research.

Several studies have provided evidence that the ability of stem cells to respond to environmental demands may be diminished during aging. Most of the studies on stem cells and aging have been performed on hematopoietic stem cells [102]. Analyses of clonogenic myeloid progenitors from bone marrow suggest that the proliferative capacity of myeloid progenitors decreases progressively with increasing age [103]. Similar results were obtained in studies in which the ability of bone marrow stromal cells to support the production of B-lymphocytes was examined [104]. Chen and coworkers used a competitive dilution assay to measure both the functional ability of hematopoietic stem cells and their concentration in BALB/cBy mice. Their data suggest that their functional ability declines during development and aging [105]. However, the stem cells in old mice exhibit an ability to repopulate engrafted host mice that is not different than stem cells from young mice. These findings suggest that, at least in the case of bone marrow, stem cell populations retain their potential to repopulate an organ throughout life. Although the proliferative potential of hematopoietic stem cells can be maintained during aging in a given species or strain of rodent, cross-species analyses suggest that the proliferative potential of hematopoietic stem cells is related to maximum lifespan [106]. Very few studies have examined the effects of aging on NPC. Incorporation of BrdU by NPC in the rat hippcampus is decreased during aging [107], an alteration which could conceivably account for age-related deficits in hippocampus-mediated brain functions such as learning and memory. Interestingly, adrenalectomy can restore neurogenesis in the hippocampus of old rats [108], demonstrating that neural stem cells in the hippocampus retain the ability for proliferation even in very old animals and suggesting a role for increased activation of the hypothalamic – pituitary – adrenal axis in age-related compromise of NPC populations.

The declining ability of the nervous system to recover from injury and disease with advancing age might result, at least in part, from a reduced capacity of NPC to respond to environmental demands. It has been shown by several laboratories that neural stem cells are mobilized in response to injury. For example, cerebral ischemia increases neurogenesis in the dentate gyrus of gerbils [72] and severe seizures increase proliferation of NPC in the same brain region of rats [70]. In addition, increased functional demands on the brain may increase the proliferation, differentiation and/or survival of NPC. Thus, the numbers of newly generated neural cells that survive are increased in the hippocampus of rodents maintained in an enriched environment [68], as well as in rats that exercise regularly [69]. It will be important to determine whether aging affects the ability of neural stem cells to respond to such environmental demands. In this regard it is of considerable interest that the survival of newly-generated cells in the dentate gyrus of the hippocampus is increased in rats maintained on a dietary restriction feeding regimen [109], a dietary manipulation known to increase lifespan.

The mechanisms that underlie changes in the ability of pluripotent NPC to proliferate, differentiate and/or survive during aging are not known. Possible players include telmomerase, growth factor and cytokine signaling pathways and oxidative stress.

Studies have shown that levels of telomerase activity increase in hematopoietic progenitor cells upon their proliferation and differentiation, and decrease with aging [110], and telomerase activity decreases greatly during the period of embryonic development when NPC cease proliferating and begin to differentiate [98, 99]. Changes in the levels of expression of some growth factors and cytokines have been documented in several tissues during normal aging and in association with age-related diseases [3], but it is not known whether these changes contribute to age-related alterations in stem cells.

Increased cellular oxidative stress is a widespread occurrence during aging and manifests as oxyradical-mediated damage to proteins, lipids and DNA. As described above, there is considerable evidence that such oxidative stress plays a central role in the aging process itself, as well as in age-related neurodegenerative disorders. Surprisingly, there is little information available on the impact of oxidative stress on stem cells. However, we have recently provided evidence for a role for nitric oxide in regulating survival of NPC (Figure 5). We found that that nitric oxide can induce death of NPC by a mechanism requiring activation of poly(ADP-ribose) polymerase and caspase-3 [111]. Nitric oxide causes release of cytochrome c from mitochondria and Bcl-2 protects the NPC against death, consistent with a pivotal role for mitochondrial changes in controlling the cell death process. These findings suggest that nitric oxide and Bcl-2 may regulate the survival of neural progenitor cells.

7.   NPC in the pathogenesis and treatment of neurodegenerative disorders

Do alterations in NPC play a role in disease initiation and/or pathogenesis? How do genetic and environmental factors impact on NPC in ways that either promote or prevent age-related disease? Can manipulations that affect endogenous NPC, or stem cell transplantation, be used to treat patients? These are the kinds of questions that are being addressed in animal models of neurodegenerative disorders with the goal of preventing the onset of the disease and/or treating patients. Age-related alterations in stem cell proliferative potential have been documented in studies of bone marrow and brain, but virtually no information on the molecular and cellular mechanisms underlying such alterations have been obtained. Nevertheless, clues are emerging from the combined efforts of molecular geneticists, developmental biologists, and molecular and cellular gerontologists. An example of one specific stem cell-regulating signaling pathway linked to several age-related diseases comes from work on Notch. In addition to regulating cell proliferation and differentiation, recent findings link Notch signaling to cell death, and suggest that alterations in Notch signaling may contribute to several age-related diseases. Overexpression of Notch can protect T lymphocytes against apoptosis induced by T cell receptor crosslinking and exposure to glucocorticoids [112]. Many apoptosis-resistant cancer cells exhibit enhanced Notch activation [113], and Notch can prevent drug-induced death of erythroleukemia cells [114]. A link between Notch and AD is suggested by the fact that the phenotype of presenilin-1 knockout mice is essentially identical to that of Notch knockout mice [74, 115], by data demonstrating that mutations in presenilin-1 can cause AD [39], and by recent findings suggesting that presenilin-1 is involved in Notch processing and/or signaling [116, 117]. Presenilin-1

Control   **Nitric Oxide**

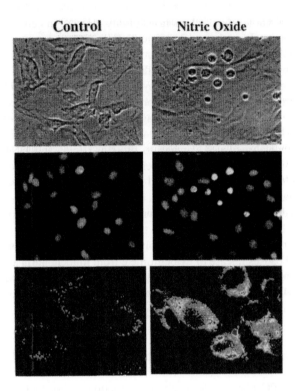

Figure 5.   Nitric oxide induces apoptosis of neural progenitor cells. Upper and middle panels) NPC were exposed to saline (Control) or 200 μM sodium nitroprusside (nitric oxide) for 7 hours. The cells were then fixed and stained with a fluorescent DNA-binding dye. The upper panels show phase-contrast images of the cells and the middle panels show nuclear fluorescence. Note that nitric oxide induced cell rounding and nuclear chromatin condensation in many of the cells. Lower panels) Images of fluorescence corresponding to sites of activated caspase-3 in NPC in a control culture and a culture that had been exposed to 200 μM sodium nitroprusside for 4 hours. Note that nitric oxide induced a marked increase in caspase-3 activity.

has been shown to modulate neurite outgrowth and cell survival of CNS neurons, and mutations in presenilin-1 that cause AD have adverse effects on neurite outgrowth and cell survival as the result of an abnormality in cellular calcium homeostasis [117-119]. Presenilin-1 mutations may also inhibit Notch signaling [116], although it remains to be determined whether inhibition of Notch signaling is required for the effects of presenilin-1 mutations on neurite outgrowth and cell survival.

Two requirements for the successful replacement of damaged neurons (or glia) by NPC are that the cells be capable of differentiating into the specific phenotype required to restore function, and that the cells find their proper position in within the damaged neural circuits. Thus, damaged dopaminergic neurons would need to be replaced in PD patients, glutamatergic or cholinergic neurons in AD, and motor neurons in ALS.

An increasing number of reports strongly suggest that NPC can indeed be coaxed into differentiating into specific subpopulations of neurons. For example, transplantation of NPC into the damaged striata in a rat model of HD resulted in many cells acquiring the phenotype of mature striatal neurons [120]. As further evidence that factors are produced in response to brain injury that stimulate NPC in a manner that promotes replacement of the specific neurons that are damaged, Nishino et al. [121] reported that mesencephalic NPC, but not cortical NPC, develop more readily into dopaminergic cells in striatum damaged by the dopaminergic toxin 6-hydroxydopamine than in uninjured striatum. However, it is not yet clear that a NPC-derived neuron can maintain its phenotype and integrate into neural circuits in the adult brain. Thus, transplantation of human NPC into the striatum of adult rodents in a model of PD resulted in an early proliferation of the cells, transient expression of tyrosine hydroxylase, and then a downregulation of tyrosine hydroxylase expression and extensive axonal outgrowth [122]. Transplantation of mouse ES cells into the spinal cord of rats that had been subjected to traumatic spinal cord injury resulted in survival, and differentiation into neurons and glial cells, of the transplanted cells [123]. Moreover, the transplanted animals showed improved hindlimb function compared to control spinal cord-injured rats. In a model of stroke in which the middle cerebral artery of rats is transiently occluded, transplantation of embryonic NPC ameliorated deficits in learning and memory caused by the ischemic brain injury [124], suggesting the possible application of stem cell therapy to human stroke patients.

Studies in animal models have shown that brain injury can induce proliferation of NPC in the hippocampus [70-72], and this response may represent an attempt to replace lost or damaged neurons. Remarkably, newly-generated neural cells can migrate long distances to a focal site of disease or injury. For example, transplanted NPC migrate rapidly to the site of active gliomas in the brains of adult rodents, even when implanted into the contralateral hemisphere [125]. Increased production and/or survival of newly generated neurons may also occur in response to physiological demands. For example, rearing of rats and mice in enriched environments results in an increase in neurogenesis [67], and maintenance of rats on a reduced calorie diet results in an increase in the survival of newly-generated hippocampal cells [109]. Increased BDNF production appears to play an important role in responses of dentate gyrus NPC to environmental enrichment and dietary restriction [109]. Thus, a better understanding of the signals that control responses of NPC to environmental demands is likely to lead to novel preventative and therapeutic approaches for neurodegenerative disorders.

The major focus of this chapter has been on neurons, because it is neurons that become dysfunctional and degenerate in the major age-related neurodegenerative disorders. However, glial cells may also become dysfunctional and degenerate in certain disorders, and changes in glia may either promote or suppress the neurodegenerative process, and it is therefore important to understand the molecular and cellular biology of glial precursor cells. There are three major types of glial cells in the central nervous system, namely, astrocytes, oligodendrocytes and microglia. NPC in the subventricular zone and dentate gyrus of the hippocampus can give rise to astrocytes [126]. Astrocytes serve a variety of functions that promote plasticity and survival of neurons including uptake of glutamate and potassium, and production of neurotrophic factors [127].

A subpopulation of NPC in the adult brain has the capacity to form oligodendrocyte-restricted precursors, which can then differentiate into mature oligodendrocytes capable of myelinating axons [128]. When such progenitor cells are transplanted into the brain or spinal cord they are capable of myelinating demyelinated axons [128, 129]. In contrast to astrocytes and oligodendrocytes, microglia arise from bone marrow progenitor cells, and are essentially macrophages that migrate into the brain [130]. Microglial activation is thought to promote neuronal degeneration in a range of disorders including AD and stroke; when activated, microglia produce toxic substances including oxyradicals and excitotoxins [131]. Suppression of microglial activation may contribute the beneficial effects of anti-inflammatory drugs that have been suggested from epidemiological and clinical studies of AD [132]. A better understanding of the impact of aging on glial cell precursors, and the molecular mechanisms regulating the proliferation, differentiation and survival of astrocytes, oligodendrocytes and microglia, may eventually reveal novel approaches for preventing and treating neurodegenerative disorders of aging.

Finally, ES cells have the potential to differentiate into any type of cell in the nervous system, and recent findings suggest that this potential might eventually be tapped in the neurology clinic. For example, it has been shown that mouse ES cells can be induced to differentiate into neurons that produce either dopamine or serotonin by treating them with certain growth factors [126]. Another potential NPC-based therapeutic approach is to use the cells as "factories" for the production of neurotrophic factors. Proof of principle of the latter approach was obtained in a study showing that NGF-secreting NPC can ameliorate ischemic damage to striatal neurons in a rodent stroke model [127]. Clearly, these are exciting times for those who study neural stem cells and their potential applications for treating neurodegenerative disorders.

8.   Acknowledgements

We thank our those who performed research in our laboratories for their important contributions to the field of stem cell research. We are also grateful for the continued support of the NIH.

9.   References

1.   Bowling, A.C. and Beal, M.F. (1995) Life Sci. 56, 1151-1171.

2.   Mattson, M.P., Pedersen, W.A., Duan, W., Culmsee, C. and Camandola, S. (1999) Ann. N.Y. Acad. Sci. 893, 154-175.

3.   Mattson, M.P. and O. Lindvall (1997) In (M.P. Mattson and J.W. Geddes, eds) The Aging Brain (JAI Press, Greenwich CT), Adv. Cell Aging Gerontol. Vol. 2, pp 299-345.

4.   Mattson, M.P. (1997) Physiol. Rev. 77, 1081-1132.

5.   Jenner, P. and Olanow, C.W. (1998) Ann. Neurol. 44, S72-84.

6.   Petersen, A., Mani, K. and Brundin, P. (1999) Exp. Neurol. 157, 1-18.

7.   Martin, L.J., Price, A.C., Kaiser, A., Shaikh, A.Y. and Liu, Z. (2000) Int. J. Mol. Med. 5, 3-13.

8.   Morrison, J.H. and Hoff, P.R. (1997) Science 278, 412-419.

9. Bertoni-Freddari, C., Fattoretti, P., Paoloni, R., Caselli, U., Galeazzi, L. and Meier-Ruge, W. (1996) Gerontology 42, 170-180, 1996.

10. Yan, S.D., Yan, S.F., Chen, X., Fu, J., Chen, M., Kupusamy, P., Smith, M.A., Perry, G., Godman, G.C., Nawroth, P. and Stern, D. (1995) Nature Med. 1, 693-699.

11. Mattson, M.P., Fu, W., Waeg, G. and Uchida, K. (1997) NeuroReport 8, 2275-2281.

12. Nixon, R.A., Saito, K.I., Grynspan, F., Griffin, W.R., Katayama, S., Honda, T., Mohan, P.S., Shea, T.B. and Beermann, M. (1994) Ann. N.Y. Acad. Sci. 747, 77-91.

13. Laping, N.J., Teter, B., Nichols, N.R., Rozovsky, I. and Finch, C.E. (1994) Brain Pathol. 4, 259-275.

14. Mattson, M.P., Culmsee, C. and Yu, Z.F. (2000) Cell Tissue Res. 301, 173-187.

15. Keller, J.N. and Mattson, M.P. (1998) Rev. Neurosci. 9, 105-116.

16. Stadtman, E.R. and Berlett, B.S. (1998) Drug Metab. Rev. 30, 225-243.

17. Schapira, A.H. (1996) Curr. Opin. Neurol. 9, 260-264.

18. Adamec, E., Vonsattel, J.P. and Nixon, R.A. (1999) Brain Res. 849, 67-77.

19. Thomas, L.B., Gates, D.J., Richfield, E.K., O'Brien, T.F., Schweitzer, J.B. and Steindler, D.A. (1995) Exp. Neurol. 133, 265-272.

20. Kruman, I.I., Culmsee, C., Chan, S.L., Kruman, Y., Guo, Z., Penix, L. and Mattson, M.P. (2000) J. Neurosci. 20, 6920-6926.

21. Disterhoft, J.F., Thompson, L.T., Moyer, J.R. Jr, and Mogul, D.J. (1996) Life Sci. 59, 413-420.

22. Mattson, M.P., LaFerla, F.M., Chan, S.L., Leissring, M.A., Shepel, P.N. and Geiger, J.D. (2000) Trends Neurosci. 23, 222-229.

23. Sapolsky R.M. (1999) Exp. Gerontol. 34, 721-732.

24. Wise, P.M., Smith, M.J., Dubal, D.B., Wilson, M.E., Krajnak, K.M. and Rosewell, K.L. (1999) Endocr. Rev. 20, 243-248.

25. Tang, M.X., Jacobs, D., Stern, Y., Marder, K., Schofield, P., Gurland, B., Andrews, H, and Mayeux, R. (1996) Lancet 348, 429-432.

26. Burgess, W., Liu, Q., Zhou, J., Tang, Q., Ozawa, A., VanHoy, R., Arkins, S., Dantzer, R. and Kelley, K.W. (1999) Neuroimmunomodulation 6, 56-68.

27. Antel, J.P. and Owens, T. (1999) J. Neuroimmunol. 100, 181-189.

28. McGeer, P.L. and McGeer, E.G. (2000) Arch. Neurol. 57, 789-790.

29. Mattson, M.P. (2000) Brain Pathol. 10, 300-312.

30. Kleim, J.A., Vij, K., Ballard, D.H. and Greenough, W.T. (1997) J. Neurosci. 17, 717-721.

31. Snowdon, D.A., Kemper, S.J., Mortimer, J.A., Greiner, L.H., Wekstein, D.R. and Markesbery, W.R. (1996) Findings from the Nun Study. JAMA 275, 528-532.

32. Prolla, T.A. and Mattson, M.P. (2001) TINS Review 24(11): S21-S31.

33. Bruce-Keller, A.J., Umberger, G., McFall, R. and Mattson, M.P. (1999) Ann. Neurol. 45, 8-15.

34. Duan, W. and Mattson, M.P. (1999) J. Neurosci. Res. 57, 195-206.

35. Yu, Z.F. and Mattson, M.P. (1999) J. Neurosci. Res. 57, 830-839.

36. Mattson, M.P. (2000) Nature Rev. Mol. Cell Biol. 1, 120-129.

37. Su, J.H., Anderson, A.J., Cummings, B. and Cotman, C.W. (1994) NeuroReport 5, 2529-2533.

38. Guo, Q., Fu, W., Xie, J., Luo, H., Sells, S.F., Geddes, J.W., Bondada, V., Rangnekar, V.M. and Mattson, M.P. (1998) Nature Med. 4, 957-962.

39. St George-Hyslop, P.H. (2000) Biol. Psychiatry 47, 183-199.

40. Mattson, M.P., Guo, Q., Furukawa, K. and Pedersen, W.A. (1998) J. Neurochem. 70, 1-14.

41. Guo, Q., Sopher, B.L., Furukawa, K., Pham, D.G., Robinson, N., Martin, G.M. and Mattson, M.P. (1997) J. Neurosci. 17, 4212-4222.

42. Guo, Q., Fu, W., Sopher, B.L., Miller, M.W., Ware, C.B., Martin, G.M. and Mattson, M.P. (1999) Nature Med. 5, 101-106.

43. Begley, J.G., Duan, W., Duff, K. and Mattson, M.P. (1999) J. Neurochem. 72, 1030-1039.

44. Betarbet, R., Sherer, T.B., MacKenzie, G., Garcia-Osuna, M., Panov, A.V. and Greenamyre, J.T. (2000) Nat. Neurosci. 3, 1301-1306.

45. de Silva, H.R., Khan, N.L. and Wood, N.W. (2000) Curr Opin Genet. Dev. 10, 292-298.

46. El-Agnaf, O.M., Jakes, R., Curran, M.D., Middleton, D., Ingenito, R., Bianchi, E., Pessi, A., Neill, D. and Wallace, A. (1998) FEBS Lett. 440, 71-75.

47. Abeliovich, A., Schmitz, Y., Farinas, I., Choi-Lundberg, D., Ho, W.H., Castillo, P.E., Shinsky, N., Verdugo, J.M., Armanini, M., Ryan, A., Hynes, M., Phillips, H., Sulzer, D. and Rosenthal, A. (2000) Neuron 25, 239-252.

48. Wellington, C.L., Singaraja, R., Ellerby, L., Savill, J., Roy, S., Leavitt, B., Cattaneo, E., Hackam, A., Sharp, A., Thornberry, N., Nicholson, D.W., Bredesen, D.E. and Hayden, M.R. (2000) J. Biol. Chem. 275, 19831-19838.

49. Ona, V.O., Li, M., Vonsattel, J.P., Andrews, L.J., Khan, S.Q., Chung, W.M., Frey, A.S., Menon, A.S., Li, X.J., Stieg, P.E., Yuan, J., Penney, J.B., Young, A.B., Cha, J.H. and Friedlander, R.M. (1999) Nature 399, 263-267.

50. Fischbek, K.H., Souders, D. and La Spada, A. (1991). Adv. Neurol. 56, 209-213.

51. Orr, H.T., Chung, M.Y., Banfi, S., Kwiatkowski, T.J., Jr., Servadio, A., Beaudet, A.L., McCall, A.E., Duvick, L.A., Ranum, L.P. and Zoghbi, H.Y. (1993). Nat. Genet. 4, 221-226.

52. Cookson, M.R. and Shaw, P.J. (1999) Brain Pathol. 9, 165-186.

53. Kostic, V., Jackson-Lewis, V., de Bilbao, F., Dubois-Dauphin, M. and Przedborski, S. (1997) Science 277, 559-562.

54. Li, M., Ona, V.O., Guegan, C., Chen, M., Jackson-Lewis, V., Andrews, L.J., Olszewski, A.J., Stieg, P.E., Lee, J.P., Przedborski, S. and Friedlander, R.M. (2000) Science 288, 335-339.

55. Kruman, I., Pedersen, W.A. and Mattson, M.P. (1999) Exp. Neurol. 160, 28-39.

56. Lipton, P. (1999) Ischemic cell death in brain neurons. Physiol. Rev. 79, 1431-568.

57. Clark, R.S., Kochanek, P.M., Chen, M., Watkins, S.C., Marion, D.W., Chen, J., Hamilton, R.L., Loeffert, J.E. and Graham, S.H. (1999) FASEB J. 13, 813-821.

58. Emery, E., Aldana, P., Bunge, M.B., Puckett, W., Srinivasan, A., Keane, R.W., Bethea, J. and Levi, A.D. (1998) J. Neurosurg. 89, 911-920.

59. Yakovlev, A.G., Knoblach, S.M., Fan, L., Fox, G.B., Goodnight, R. and Faden, A.I. (1997) J. Neurosci. 17, 7415-7424.

60. Pedersen, R.A. (1994) Studies of in vitro differentiation with embryonic stem cells. Reprod. Fertil. Dev. 6, 543-552.

61. Okabe, S., Forsberg-Nilsson, K., Spiro, A.C., Segal, M. and McKay, R.D. (1996) Mech. Dev. 59, 89-102.

62. Armstrong, L., Lako, M., Lincoln, J., Cairns, P.M. and Hole, N. (2000) Mech. Dev. 97, 109-116.

63. Gage, F.H. (2000) Science 287, 1433-1438.

64. Mujtaba, T., Piper, D.R., Kalyani, A., Groves, A.K., Lucero, M.T. and Rao, M.S. (1999) Dev. Biol. 214, 113-127.

65. Rao, M.S. (1999) Anat. Rec. 257, 137-148.

66. Temple, S. (2000). Humana Press. (Rao, M.S., editor). In press.

67. Kempermann, G., Kuhn, H.G., Gage, F.H. (1997) Nature 386, 493–495.

68. Young, D., Lawlor, P.A., Leone, P., Dragunow, M., and During, M.J. (1999) Nature Med. 5, 448-453.

69. Neeper, S.A., Gomez-Pinilla, F., Choi, J., and Cotman, C.W. (1996) Brain Res. 726, 49-56.

70. Parent, J.M., Yu, T.W., Leibowitz, R.T., Geschwind, D.H., Sloviter, R.S., and Lowenstein, D.H. (1997) J. Neurosci. 17, 3727-3738.

71. Tzeng, S.F., and Wu, J.P. (1999) Neuroreport 10, 2287-2292.

72. Liu, J., Solway, K., Messing, R.O. and Sharp, F.R. (1998) J. Neurosci. 18, 7768-7778.

73. Artavanis-Tsakonis, S., Matsuno, K. and Fortini, M.E. (1995) Science 268, 225-231.

74. Conlon, R.A., Reaume, A.G. and Rossant, J. (1995) Development 121, 1533-1542.

75. Guan, E., Wang, J., Laborda, J., Norcross, M., Baeuerle, P.A. and Hoffman, T. (1996) J. Exp. Med. 183, 2025-2032.

76. Heitzler, P., and Simpson, P. (1991) Cell 64, 1083-1091.

77. Guo, M., Jan, L.Y., and Jan, Y.N. (1996) Neuron 17, 27-41.

78. Verdi, J.M., Bashirullah, A., Goldhawk, D.E., Kubu, C.J., Jamali, M., Meakin, S.O. and Lipshitz, H.D. (1999) Proc. Natl. Acad. Sci. USA 96, 10472-10476.

79. Zhong, W., Jiang, M.M., Schonemann, M.D., Meneses, J.J., Pedersen, R.A., Jan, L.Y. and Jan, Y.N. (2000) Proc. Natl. Acad. Sci. USA 97, 6844-6849.

80. Verdi, J.M., Bashirullah, A., Goldhawk, D.E., Kubu, C.J., Jamali, M., Meakin, S.O. and Lipshitz, H.D. (1999) Proc. Natl. Acad. Sci. USA 96, 10472-10476.

81. Gritti, A., Frolichsthal-Schoeller, P., Galli, R., Parati, E.A., Cova, L., Pagano, S.F., Bjornson, C.R. and Vescovi, A.L. (1999) J. Neurosci. 19, 3287-3297.

82. Eaton, M.J., and Whittemore, S.R. (1996) Exp. Neurol. 140, 105-114.

83. Arsenijevic, Y. and Weiss, S. (1998) J. Neurosci. 18, 2118-2128.

84. Gary, D.S. and Mattson, M.P. (2001) J. Neurochem. 76, 1485-1496.

85. Jacques, T.S., Relvas, J.B., Nishimura, S., Pytela, R., Edwards, G.M., Streuli, C.H. and ffrench-Constant, C. (1998) Development 125, 3167–3177.

86. Palmer, T.D., Markakis, E.A., Willhoite, A.R., Safar, F. and Gage, F.H. (1999) J. Neurosci. 19, 8487-8497.

87. Whittemore, S.R., Morassutti, D.J., Walters, W.M., Liu, R.H. and Magnuson, D.S. (1999) xp. Cell Res. 252. 75-95.

88. Pincus, D.W., Keyoung, H.M., Harrison-Restelli, C., Goodman, R.R., Fraser, R.A., Edgar, M., Sakakibara, S., Okano, H., Nedergaard, M. and Goldman, S.A. (1998) Ann. Neurol. 43, 576-585.

89. Hayashi, Y., Kashiwagi, K., Ohta, J., Nakajima, M., Kawashima, T. and Yoshikawa, K. (1994) Biochem. Biophys. Res. Commun. 205, 936-943.

90. Mattson, M.P., Cheng, B., Culwell, A., Esch, F., Lieberburg, I. and Rydel, R.E. Cheng, A. Culwell, F. Esch, I. Lieberburg and R.E. Rydel (1993) Neuron 10, 243-254.

91. Furukawa, K., Barger, S.W., Blalock, E. and Mattson, M.P. (1996) Nature 379, 74-78.

92. Taupin, P., Ray, J., Fischer, W.H., Suhr, S.T., Hakansson, K., Grubb, A. and Gage, F.H. (2000) Neuron 28, 385-397.

93. Ma, W., Maric, D., Li, B.S., Hu, Q., Andreadis, J.D., Grant, G.M., Liu, Q.Y., Shaffer, K.M., Chang, Y.H., Zhang, L., Pancrazio, J.J., Pant, H.C., Stenger, D.A. and Barker, J.L. (2000) Eur. J. Neurosci. 12, 1227-1240.

94. Pende, M., Fisher, T.L., Simpson, P.B., Russell, J.T., Blenis, J. and Gallo, V. (1997) J. Neurosci. 17, 1291-1301.

95. Mattson, M.P. (1988) Brain Res. Rev. 13, 179-212.

96. Mattson, M.P. and Klapper, W. (2001) J. Neurosci. Res. 63, 1-9.

97. Fu, W., Begley, J.G., Killen, M.W. and Mattson, M.P. (1999) J. Biol. Chem. 274, 7264-7271.

98. Fu, W., Killen, M., Pandita, T. and Mattson, M.P. (2000) J. Mol. Neurosci. 14, 3-15.

99. Klapper, W., Shin, T. and Mattson, M.P. (2001) J. Neurosci. Res. 64, 252-260.

100. Evans, S.K., Sistrunk, M.L., Nugent, C.I. and Lundblad, V. (1998) Chromosoma 107, 352-358.

101. d'Adda di Fagagna, F., Hande, M.P., Tong, W.M., Lansdorp, P.M., Wang, Z.Q. and Jackson, S.P. (1999) Nat. Genet. 23, 76-80.

102. Globerson, A. (1999) Hematopoietic stem cells and aging. Exp. Gerontol. 34, 137-146.

103. Marley, S.B., Lewis, J.L., Davidson, R.J., Roberts, I.A., Dokal, I., Goldman, J.M. and Gordon, M.Y. (1999) Br. J. Haematol. 106, 162-166.

104. Stephan, R.P., Reilly, C.R. and Witte, P.L. (1998) Blood 91, 75-88.

105. Chen, J., Astle, C.M. and Harrison, D.E. (1999) Exp. Hematol. 27, 928-935.

106. Van Zant, G. (2000) Results Probl. Cell Differ. 29, 203-235.

107. Kuhn, H.G., Dickinson-Anson, H., and Gage, F.H. (1996) J. Neurosci. 16, 2027-2033.

108. Montaron, M.F., Petry, K.G., Rodriguez, J.J., Marinelli, M., Aurousseau, C., Rougon, G., Le Moal, M. and Abrous, D.N. (1999) Eur. J. Neurosci. 11, 1479-1485.

109. Lee, J., Duan, W., Long, J.M., Ingram, D.K. and Mattson, M.P. (2000) J. Mol. Neurosci. 15, 105-113.

110. Hiyama, K., Hirai, Y., Kyoizumi, S., Akiyama, M., Hiyama, E., Piatyszek, M.A., Shay, J.W., Ishioka, S. and Yamakido, M. (1995) J. Immunol. 155, 3711-3715.

111. Cheng, A. Chan, S.L., Milhavet, O., Wang, S. and Mattson, M.P. (2001) J. Biol. Chem. 276, 43320-43327.

112. Jehn, B.M., Bielke, W., Pear, W.S. and Osborne, B.A. (1999) J. Immunol. 162, 635-638.

113. Jeffries, S. and Capobianco, A.J. (2000) Mol. Cell. Biol. 20, 3928-3941.

114. Shelly, L.L., Fuchs, C., and Miele, L. (1999) J. Cell Biochem. 73, 164-175.

115. Shen, J., Bronson, R.T., Chen, D.F., Xia, W., Selkoe, D.J. and Tonegawa, S. (1997) Cell 89, 629-639.

116. Song, W., Nadeau, P., Yuan, M., Yang, X., Shen, J. and Yankner, B.A. (1999) Proc. Natl. Acad. Sci. USA 96, 6959-6963.

117. Handler, M., Yang, X. and Shen, J. (2000) Development 127, 2593-2606.

118. Furukawa, K., Guo, Q., Schellenberg, G.D. and Mattson, M.P. (1998) J. Neurosci. Res. 52, 618-624.

119. Guo, Q., Sebastian, L., Sopher, B.L., Miller, M.W., Glazner, G.W., Ware, C.B., Martin, G.M. and Mattson, M.P. (1999) Natl. Acad. Sci. USA 96, 4125-4130.

120. Armstrong, R.J., Watts, C., Svendsen, C.N., Dunnett, S.B. and Rosser, A.E. (2000) Cell Transplant. 9, 55-64.

121. Nishino, H., Hida, H., Takei, N., Kumazaki, M., Nakajima, K. and Baba, H. (2000) Exp. Neurol. 164, 209-214.

122. Ostenfeld, T., Caldwell, M.A., Prowse, K.R., Linskens, M.H., Jauniaux, E. and Svendsen, C.N. (2000) Exp. Neurol. 164, 215-226.

123. McDonald, J.W., Liu, X.Z., Qu, Y., Liu, S., Mickey, S.K., Turetsky, D., Gottlieb, D.I. and Choi, D.W. (1999) Nature Med. 5, 1410-1412.

124. Fukunaga, A., Uchida, K., Hara, K., Kuroshima, Y., and Kawase, T. (1999) Cell Transplant. 8, 435-441.

125. Aboody, K.S., Brown, A., Rainov, N.G., Bower, K.A., Liu, S., Yang, W., Small, J.E., Herrlinger, U., Ourednik, V., Black, P.M., Breakefield, X.O. and Snyder, E.Y. (2000) Proc. Natl. Acad. Sci. USA 97, 12846-12851.

126. Nilsson, M., Perfilieva, E., Johansson, U., Orwar, O. and Eriksson, P.S. (1999) J. Neurobiol. 39, 569-578.

127. Muller, H.W., Junghans, U. and Kappler, J. (1995) Pharmacol. Ther. 65, 1-18.

128. Zhang, S.C., Ge, B. and Duncan, I.D. (1999) Proc. Natl. Acad. Sci. USA 96, 4089-4094.

129. Bulte, J.W., Zhang, S., van Gelderen, P., Herynek, V., Jordan, E.K., Duncan, I.D. and Frank, J.A. (1999) Proc. Natl. Acad. Sci. USA 96, 15256-15261.

130. Mattson, M.P. (2001) Trends Neurosci. 24, 5.

131. Zielasek, J. and Hartung, H.P. (1996) Adv. Neuroimmunol. 6, 191-122.

132. Flynn, B.L. and Theesen, K.A. (1999) Ann. Pharmacother. 33, 840-849.

133. Lee, S.H., Lumelsky, N., Studer, L., Auerbach, J.M. and McKay, R.D. (2000) Nat. Biotechnol. 18, 675-679.

134. Andsberg, G., Kokaia, Z., Bjorklund, A., Lindvall, O. and Martinez-Serrano, A. (1998) Eur. J. Neurosci. 10, 2026-2036.

135. Siegel, G.J. and Chauhan, N.B. (2000) Brain Res. Rev. 33, 199-227.

136. Galli, R., Pagano, S.F., Gritti, A. and Vescovi, A.L. (2000) Dev. Neurosci. 22, 86-95.

137. Suzuki, S., Tanaka, K., Nogawa, S., Ito, D., Dembo, T., Kosakai, A. and Fukuuchi, Y. (2000) J.Cereb. Blood Flow Metab. 20, 661-668.

138. Kiss, J.Z. (1998) Mol. Cell. Endocrinol. 140, 89-94.

139. Mikkonen, M., Soininen, H., Tapiola, T., Alafuzoff, I. and Miettinen, R. (1999) Eur. J. Neurosci. 11, 1754-1764.

140. Haydar, T.F., Wang, F., Schwartz, M.L. and Rakic, P. (2000) J Neurosci. 20, 5764-5774.

141. Haughey, N.J., Culmsee, C. and Mattson, M.P. (2000) Soc. Neurosci. Abstr. 26, 1784.

Table 1. Signaling mechanisms that affect neural progenitor cells that may be altered in the nervous system during aging and in age-related neurodegenerative disorders.

Factor	Effect on NPC	Change in aging or disease	References
bFGF	IP, ID, IS	reactive increase; decrease in PD	3, 63
BDNF	ID, IS	decrease in aging and AD	82, 109, 135
LIF	IP, DD	increase in stroke	136, 137
NCAM	ID	aberrant expression in AD	138, 139
Glutamate	IP, DP	facilitates death in AD, PD, stroke	2, 140
Nitric oxide	DS	increase in AD and stroke	56, 111
Amyloid	DP, DS	increased deposition in brain	141

I, increased; D, decreased; P, proliferation; D, differentiation; S, survival.

Table 2. Genetic and environmental factors in age-related neurodegenerative disorders.

Disorder	Genetic factors	Environmental factors
Alzheimer's disease	APP and presenilin mutations	high calorie diet; head trauma
	ApoE polymorphism	low education
Parkinson's disease	α-synuclein, Parkin mutations	high calorie intake; pesticides
Huntington's disease	huntingtin trinucleotide expansions	none yet established
ALS	Cu/Zn-SOD mutations	toxins
Stroke	Notch-3 mutations	

CHAPTER 7

MYOCARDIAL AGING AND EMBRYONIC STEM CELL BIOLOGY

KENNETH R. BOHELER and ANNA M. WOBUS

Table of contents

---

1.   Aging and cardiovascular disease

The complex process of aging progressively damages organ systems [1]; however, the past several centuries have revealed how medical and social advances have

Stem Cells: A Cellular Fountain of Youth.   Ed. by Mark P. Mattson and Gary Van Zant. 141 — 176

led to major increases in life expectancy at all ages. These increases have been accompanied by a much higher incidence of aging-associated disorders seen as cancer, diabetes, Alzheimer's disease, and cardiovascular diseases [2-11]. Cardiovascular diseases continue to be the major causes of morbidity and mortality (atherosclerosis, hypertension, stroke and heart failure) in developed countries. The frequency of cardiovascular disease increases exponentially in older persons and reaches epidemic proportions in the very old, affecting 10% or more of persons ≥80 years of age [10, 12]. Fewer than 30% of elderly individuals survive 6 years after their first hospitalization for heart failure [10]. As access to medicine improves and the global median age continues to increase, the impending epidemic crisis and dilemma of how to treat cardiovascular diseases globally and cost effectively presents an enormous challenge for biomedical researchers and practitioners.

The consequences of aging, disease and life-style on the heart are profound. The heart mass increases, on average, from age 30 to 90 years by 1 g/year in men and 1.5 g/year in women; however, left ventricular mass often decreases in the very old (80-100 years of age) [12]. This decrease in mass is at least partially due to a loss of cardiac myocytes. In turn, the reorganization of heart muscle provokes a whole body response to diminished cardiac function [13]. The major changes affecting cardiac output with aging are decreased heart rates and adaptations in response to arterial stiffening. When the limits of adaptation are reached, the result is often heart failure. Short-term, the adaptations are beneficial, but long-term, asymptomatic left ventricular dysfunction often progresses to symptomatic heart failure. Therapeutic approaches for treating heart failure (pharmacologic, use of mechanical devices and surgical interventions) are tailored to limit progression of the disease(s), reduce adverse effects on the patient, and maintain quality of life [13]. These interventions have little or no effect on the complex process of aging. Current therapies are not always applicable to every patient, and novel approaches are required to ultimately treat the specific and underlying molecular causes responsible for aging and progression of cardiovascular diseases in individual patients [13].

Genes exert strong controls over lifespan and patterns of aging [14], but the role of individual genes in this process is clearly affected by environmental factors. One of the challenges of aging research is to determine which genetic and environmental interactions are implicated. Ultimately, the goal will be to take appropriate steps (medical or life-style) in individuals to limit the negative impacts of aging, life-style and disease. Because progress in human aging research has largely been empirical, a more thorough molecular analysis is required before clinical benefits can be delivered to patients. Until this is achieved or where simple cost effective interventions fail, alternative strategies and therapies are required to reduce the impact of aging on morbidity and mortality of individuals.

2.   Exogenous stem cells and cell replacement therapy

Quiescent stem cells or progenitor cells promote organ or tissue self-repair. Where self-repairs are insufficient, tissue damage is extensive, or cell number and function are markedly reduced, as is frequently the case in disease, alternative therapies in the form of transplantation or cell replacement may be indicated. This is particularly true for the heart, an organ largely incapable of replacing lost or damaged muscle cells because it has little or no capacity for hyperplastic growth [15, 16] (Figure 1).

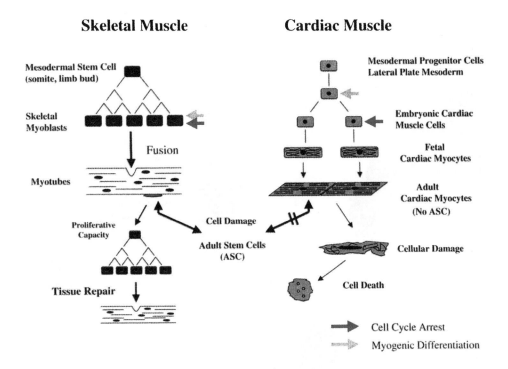

Figure 1. Differentiation and regenerative capacity of two types of muscle cells. Skeletal muscle cells are derived from mesodermal stem cells originating in somites or limb buds. Committed cells give rise to skeletal muscle myoblasts capable of active proliferation and myogenic differentiation. Upon fusion, myotubes form and lose the capacity to proliferate. A small percentage of myogenic adult stem cells (ASC) populate skeletal muscle, which in response to stress, cell damage or cell loss, are capable of proliferating and eventually replacing or repairing damaged skeletal muscle. In contrast, cardiac muscle cells are derived from progenitor cells originating in the lateral plate mesoderm. These cells have a high proliferative capacity, and as the organ forms, the cells differentiate into mature cell types, capable of contracting and relaxing. In development, cells of the compact zone in the developing myocardium rapidly divide, while those of the trabecular regions form working myocardium with more limited proliferative capacity [50]. In mammals and shortly after birth, most myocardial cells lose the capacity to divide. There are no known cardiac adult stem cells, and in response to stress, cell damage or cell loss, the myocardium is limited in its capacity to repair itself. The primary response is cardiac hypertrophy, but if the damage is very extensive, the heart muscle may ultimately fail to meet the functional requirements of the organism, the results of which are catastrophic.

Interventions currently available to treat damaged heart muscle include changes in life-style, medications that reduce functional requirements, or surgical interventions that improve cardiac function. In severe cases, transplantation is the only option. Today's most urgent problem in transplantation, however, is the lack of suitable donor organs and tissues. In fact as few as 5% of the organs needed in the US will be available for those awaiting transplantation [17]. As the population ages and heart disease becomes more prevalent, demands for organs and tissue therapies will only increase. An alternative to heart transplantation is cell therapy whose aim it is to replace, repair or enhance the biological function of damaged tissue or diseased organs [15, 16, 18] (Figure 2). One goal of cardiac cellular transplantation has been to find a renewable source of cells that can be used in human hearts [19].

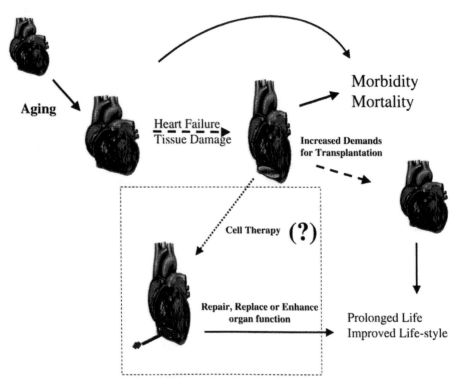

Figure 2.   The absolute mass of the adult myocardium increases with aging. With senescence, the frequency of heart failure due to myocyte loss or tissue damage increases exponentially. When therapeutic interventions are insufficient, cardiac transplantation is often the only viable clinical option to prolong life and improve life-style of patients with severe heart disease. Unfortunately, the availability of donor hearts (organs) is very limited, while the demands for transplantation continue to grow. Cell therapy has emerged as a potential intervention for the treatment of many diseases, but the absence of cardiac adult stem cells has prevented their use in heart. The aim has thus been to identify cells or factors that could activate proliferation of cells (see ? in figure) that could serve as a source of transplantable cardiac myocytes that might function in cell therapy treatments (see boxed area). Until this is achieved, the incidences of morbidity and mortality will continue to increase.

Cell replacement therapy is potentially achievable by the transfer of isolated and defined cell populations to a target organ in sufficient numbers and quality for them to survive and restore function. Sources of cells or tissues are self (autologous), same species (allogeneic), different species (xenographic), primary or immortalized (cell lines) and embryonic and adult stem cell-derived donor cells. The ability to cultivate, multiply and manipulate these cell types has either limited or encouraged their use in specific treatment protocols [18]. Where possible, autologous cells are preferential, but frequently, as in heart, unavailable (see Figure 1). Cell lines that are self-renewing and pluripotent open the possibility for cell therapy. In mammals, the property of pluripotentiality or the ability to develop to many cell types is limited to the oocyte, the zygote, early embryonic cells, primordial germ cells and embryonic carcinoma (EC) cells derived from embryo-derived tumors. The recent descriptions of human embryonic stem (ES) and embryonic germ (EG) cells herald the eventual use of autologous or allogeneic *in vitro*-differentiated cells for human therapy [20-22]. Theoretically, human ES cell lines can be maintained indefinitely *in vitro* in a pluripotent state, induced to differentiate to specific cell types that can be used to isolate pure cell populations, and eventually, homogeneous populations used in cell therapies. To be realized, however, human stem cells must exhibit the full proliferative potential and *in vitro* differentiation capacity of mouse ES cells [23, 24] (Table 1).

Table 1.   Characteristics of pluripotent embryonic Stem cell lines: mouse *versus* human (data according to [20,143]).

Characteristic	Mouse ES cells	Human ES cells
Unlimited, undifferentiated proliferation	Yes	[1]Probable
Compact, multilayered colonies	Yes	Yes
High nuclear to cytoplasmic ratio	Yes	Yes
Alkaline phosphatase activity	Yes	Yes
Stage Specific Embryonic Antigens	SSEA-1	SSEA-3, -4
Proteoglycans	No	TRA-1-60, TRA-1-8 1, GCTM-2
High activity of telomerase	Yes	Yes
Stable developmental potential	Yes	Possible
Normal and stable karyotype	Yes	Yes
Feeder cell dependence	Yes (or LIF)	Yes
Stem cell renewal factor(s)	IL-6 cytokine family	Unknown
Express Oct-4	Yes	Yes
Short Gl phase of cell cycle	Yes	Unknown
Potential for in vitro differentiation	Yes	Yes
Differentiate into cells of all three germ layers	Yes	Yes
Ability to contribute to germline	Yes	Unknown/unethical

[1] Not demonstrated at present

Extensive knowledge of mouse and human stem cell biology is required before any clinical benefits can be delivered. Of paramount importance are the questions of self-renewal, specification, control of differentiation, and how pure populations of differentiated cells, which have been grown and genetically manipulated in culture, interact in the organ after transplantation, and adapt with aging. Once a thorough understanding of the behavior of human ES cells in animal models is achieved, a future for human ES cells in clinical settings may be credible. Such a source of cells may ultimately replace the need for long-term medical treatments or heart transplantation. This goal is laden with obstacles that will require integration of cell biology, immunology, molecular biology, tissue engineering, materials science, transplantation biology and the clinical expertise before any eventual interventions can proceed [19].

3.   Mouse embryonic stem cells

3.1.   A versatile model system

Much of what we know about ES cells has been derived from studies on mouse ES cell lines. Mouse ES cells are continuous cell lines derived from the inner cell mass or epiblast [25, 26] (Figure 3). ES cells, when grown on feeder layers of primary mouse embryonic fibroblasts or leukemia inhibitory factor-producing mouse fibroblast cell lines (STO cells), have apparently an unlimited potential to propagate *in vitro* in an undifferentiated state [24, 27]. *In vitro*, these cells can maintain a relatively normal and stable karyotype, even with continual passaging. Mouse ES cells are naturally pluripotent, not totipotent, as they do not normally generate the trophoblast, the region that gives rise to the placenta, umbilical cord and amnion [28, 29]. Mouse ES cells are, however, capable of developing into almost any other cell type of the embryo. Upon removal of the mouse embryonic fibroblast feeder layer or, more generally, in the absence of leukemia inhibitory factor, ES cells spontaneously differentiate into derivatives of all three primary germ layers: endoderm, ectoderm and mesoderm (Figure 3) [30-32]. The *in vitro* differentiation of ES cells normally requires an initial aggregation to form structures, termed embryoid bodies, which then differentiate into a variety of specialized cell types, including cardiac myocytes (atrial, ventricular, pacemaker), smooth and skeletal muscle cells, hematopoietic cells, cartilage, melanocytes and neurons [33] (Figure 3). Differentiation of ES cells and formation of embryoid bodies have provided a major opportunity to study the regulation of lineage commitment, and the identification of novel precursor cells [34]. When injected into mice, ES cells also form teratomas that contain cell types derived from all three germ layers [32, 35]. When returned to the embryonic environment by transfer into a host blastocyst or aggregation with blastomere stage embryos, ES cells behave like normal embryonic cells. They can contribute to all tissues in the resulting chimeras, including germ cells. ES cells thus can be reincorporated into normal embryonic development and have the full potential to develop along all lineages of the embryo proper [36].

Mouse ES cells are amenable to genetic manipulation through random insertion or targeting events [37], and the mutations introduced into ES cells can also be transmitted

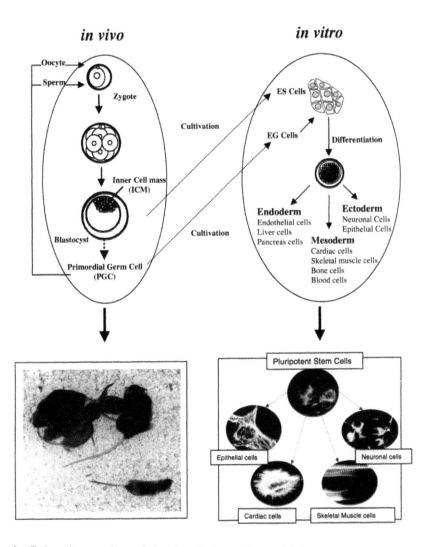

Figure 3. Embryonic stem cells are derived from the inner cell mass (ICM) and epiblast. The segregation of the germline in mouse embryos takes place in an extraembryonic region where cell contacts and local signals induce differentiation of some epiblast cells to primordial germ cells (PGC). Following appropriate *in vivo* developmental pathways, these cells contribute to all cells of a developing embryo, leading to the generation of a mature organism. ES cells can also be genetically manipulated in culture, and injected into the blastocyst where they contribute to all embryonic lineages, including the germline of chimeric embryos. Due to this feature, these cells are useful in the generation of mice carrying gene mutations whose function can be studied *in vivo*. Both ES cells and EG cells can be cultured and studied *in vitro*. When allowed to form cell aggregates, or embryoid bodies, these cells spontaneously differentiate into cells typical of all three primary germ layers. Immunofluorescence assay of undifferentiated PG cells (SSEA-1 mAb), differentiated epithelial cells (TROMA-1 mAb), differentiated cardiac and skeletal muscle cells (anti-titin T12 mAb), and neuronal cells (anti-neurofilament 160 kDa, NN18 mAb) are indicative of cells that can be isolated and studied by *in vitro* assays (Adapted from [33, 164]). PG- and ES-derived cells, *in vitro*, are useful for the study of embryogenesis, abnormal development, functional genomics, pharmacotoxicology (drug screening), and embryotoxicity of potential teratogens.

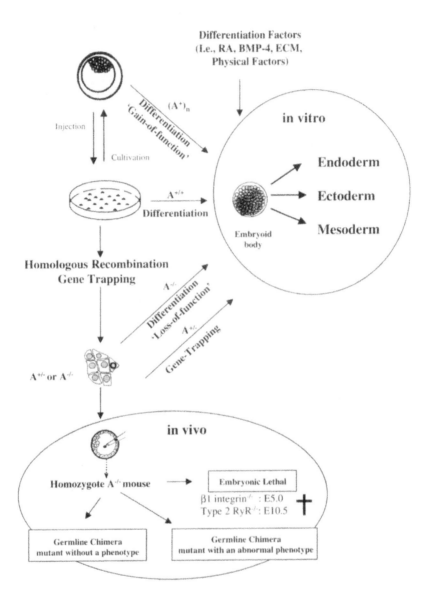

Figure 4.   Embryonic stem cell technologies (Adapted from [42]). Modulation of embryoic development by differentiation factors, 'gain-of-function', 'loss-of-function' or gene-trapping strategies *in vitro*. Pluripotent ES cells derived from mouse blastocysts and differentiated as embryoid bodies develop into cells of endodermal, ectodermal and mesodermal lineages. The *in vitro* differentiation to these cells is influenced by a number of factors. These include the number of cells involved in forming the embryoid body, the addition or removal of differentiation factors like retinoic acid (RA), bone morphogenetic protein-4 (BMP-4), or the presence of proteins from the extracellular matrix (ECM). ES cells can be genetically manipulated. When DNA is transferred into ES cells that promote constitutive expression of a gene product (A$^+$), the potential effects of the gene(s) can be examined during *in vitro* differentiation.

to the germline through the production of mouse chimeras (Figure 4). Random insertion into the mammalian genome of expression vectors permits 'gain-of-function' studies [38, 39]. These strategies permit the study of specific gene over-expression on a given differentiation lineage without having to worry about the effects of over-expression on the overall embryonic development [40]. This has, for example, permitted a more thorough characterization of specific developmental pathways such as those associated with expression of myogenic factors during skeletal muscle differentiation [41]. Random insertion of DNA into single sites (gene trapping) in the genome additionally provides a strategy for functional genomics that has led to the identification of specific factors important for commitment and differentiation. ES cells have also been used to understand the role of genes through 'loss-of-function' studies (Figure 4). Using this strategy, ES cell lines that lack a specific gene can be investigated [42-44]. This is particularly of interest *in vitro* when the inactivation of a gene is lethal in the animal. Applying gene targeting technology to ES cells in culture thus allows genetic modification of endogenous genes and study of their function both *in vitro* and *in vivo*. ES cells represent capture in culture of the normally transient phase of embryonic pluripotentiality [45], and permit gene targeting to study the effects of genes both *in vitro* and *in vivo*.

3.2.   Developmental studies of ES cells

Development is a complex biological process requiring the integration of cell specification, differentiation, and morphogenesis. The process of development is controlled by regulatory genes acting as genetic switches to stimulate signaling cascades that control gene expression, producing tissue and organismal phenotypes. In Drosophila, for example, these regulatory genes are often transcription factors [46-48]. These factors initiate gene cascades that result in phenotypic changes at each embryonic stage and direct expression of other regulators important for subsequent stages of embryonic development. Mammalian heart development is composed of two major states: early specification and differentation of cardiac myocytes and later morphogenesis. The embryonic heart is difficult to manipulate and there are no appropriate cardiac cell lines that can be used to study commitment into the cardiogenic lineage as well as study early cardiac development (for further information see Table 1 in [49]) [50]. The temporary regulatory cascades involved in expression of genes is therefore less well understood, but it is likely that mechanisms similar to those found in Drosophila exist where functional regulators play important roles in lineage commitment and stage

Caption to Figure 4 - continued.   Similarly, ES cells that have been targeted to inactivate genes by homologous recombination, resulting in a heterozygous mutation ($A^{+/-}$) or a homozygous inactivation ($A^{-/-}$), can be studied *in vitro*. Successful gene trapping strategies usually generate a heterozygous mutation ($A^{+/-}$) that activates a reporter gene. Expression of this reporter, following *in vitro* differentiation, indicates the functional capacity of the gene in specific cells. ES cells manipulated *in vitro* can also be injected into blastocysts. The cells colonize the ICM and after transplantation of the manipulated blastocysts into pseudopregnant foster mothers, chimeric animals can be generated for *in vivo* studies. In those instances where gene targeting results in early embryonic death (e.g., β1 integrin and RyR2), the *in vitro* loss-of-function strategy is an alternative to analyze the effects of that specific mutation on cellular differentiation.

progression [46, 48, 51]. In mouse, the process is clearly dynamic, and involves not only transient and long-term gene activation events, but also gene repression events [52, 53]. ES cells represent a model system to study developmental paradigms of virtually any cell type.

3.2.1. Why study development with respect to cardiac aging and disease?

Cardiac aging and disease are characterized by re-expression of gene products and signaling cascades normally associated with fetal cardiac development. These phenomena have been described as the induction of a 'fetal type gene program' [50, 54-56]. If gene expression pathways activated in aging and disease are truly representative of a reactivated fetal gene program, then delineation of fetal pathways should provide clearer insights into the adaptive mechanisms of aging and disease. Hence the reason for studying development vis-à-vis aging. Examination of cardiac development may provide essential information about potential regulatory mechanisms involved in cardiac growth, disease and senescence (including apoptosis). Identification of those signals that lock cardiac myocytes out of the cell cycle might even enable a therapy whereby cell cycle checkpoints can be bypassed, thus reinitiating cell growth in post-mitotic cardiac cells [57]. Alternatively, by identifying the regulatory factors that specify the cardiac phenotype, it may be possible to convert non-cardiac cells (e.g., fibroblasts, hematopoietic stem cells) into cardiac myocytes [58]. Transformed cells could then be used to replace damaged myocardial tissues [15].

ES cells can be maintained indefinitely *in vitro*, induced to differentiate to specific cell types, including cardiac myocytes, and represent a potential source of cardiac cells for cell therapy. Examination of the developmental potential of ES cells differentiating to cardiac myocytes is therefore critical to determine signaling pathways involved in development of cardiac myocytes and, potentially implicated in aging and disease. Following genetic manipulation, they are also valuable tools to determine the function of specific genes and the effects of these genes on differentiation to specific cell types.

3.2.2. Signaling pathways involved in self-renewal and differentiation

Embryonic stem cell self-renewal depends on paracrine signals. ES cells initially were cultured only on monolayers of inactivated fibroblasts from mouse embryos and EC cell-supplemented media [30-33]. Without these fibroblasts, pluripotentiality could not be maintained, suggesting that either direct interactions with or secreted products derived from fibroblasts promoted self-renewal. Subsequent isolation of the cytokine called leukemia inhibitory factor (LIF) or differentiation inhibiting activity (DIA) established that ES cell self-renewal, in fact, was dependent upon paracrine signal(s) [24, 27, 60]. LIF is a member of a family of related cytokines (IL-6, IL-11, OSM, CNTF, CT1) that act through a receptor complex composed of a low affinity LIF receptor (LIFR) and gp130 receptor molecules [61]. Gp130 receptor complexes activate a wide range of effector molecules. These include signal transducer and activator of transcription (STAT) 1, 3 and 5 transcription factors, insulin receptor substrate (IRS) proteins, the tyrosine phosphatase SHP-2, mitogen-activated protein kinases (MAPK), extracellular regulated kinases (ERK) 1 and 2, phosphoinositol-3 kinase and the Src family tyrosine kinases, Hck, Btk and Fes. Neither the LIFR nor the gp130 molecule has any intrinsic protein kinase domain, but each associates constitutively with the JAK family of non-receptor cytoplasmic protein tyrosine kinases. Binding of LIF to the receptor complex rapidly activates JAK tyrosine kinases, and potentially other non-receptor tyrosine kinases, leading to phosphorylation of the LIFR and gp130 on tyrosine residues. In response to phosphorylation, SH2-domain containing molecules (e.g., STATs and SHP-2) are recruited to the receptor complex and subsequently phosphorylated on their tyrosine residues by JAK tyrosine kinases. Upon phosphoryla- tion, STATs dimerize, translocate to the nucleus, and bind to DNA to direct gene(s) transcription. Based on results from chimeric receptors [62], STAT3 activation alone via gp130 is required and sufficient to maintain the undifferentiated state of ES cells.

Based on numerous studies (for review see [61]), STAT3 appears to play a critical role in mediating gp130 dependent self-renewal signals in ES cells (Figure 5). It is, however, unclear how STAT3 promotes self-renewal. It is possible that STAT3 either maintains or represses the expression of specific genes that define a pluripotent phenotype. Several candidate genes and pathways have been proposed, including

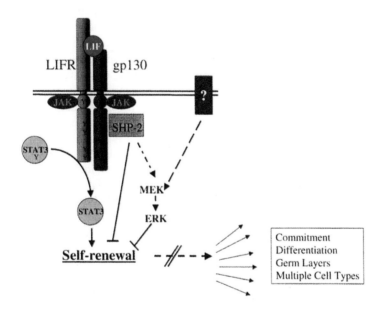

Figure 5.   Parcrine signals are necessary for self-renewal. LIF (IL-6 family of cytokines) binds to the LIFR/gp130 signaling complex. Binding of LIF to the receptor complex activates the receptor associated JAK tyrosine kinases, leading to phosphorylation of LIFR and gp130. In response, SH2-domain containing molecules (e.g., STATs and SHP-2) are recruited to the receptor complex and phosphorylated. This stimulates both STAT and MAPK signaling pathways, and interactions of these signaling pathways with other unknown pathways are sufficient to maintain the undifferentiated state of ES cells and promote self-renewal. Removal of LIF prevents phosphorylation of the receptor complex, leading to a rapid cascade of events promoting differentiation. [Adapted from [61])

maintenance of Oct-4 gene expression [63], regulation of Fos, Jun and Bcl-2 expression [64], blockade of ERK signaling pathways, or interactions that regulate the cell cycle of ES cells [61]. Oct-4 in mouse is a POU domain transcription factor with expression limited to pluripotent cells [65-67]. It is believed that the transcription factor Oct-4 is essential for establishment of the pluripotent stem cell population of the inner cell mass; whereas, Fos, Jun and ERK signaling pathways are involved in multiple facets of cellular signaling, many of which are not involved in self-renewal. There is also considerable evidence suggesting that other non-gp130 signals or other STAT family members are involved in self-renewal [61, 68, 69]. Clearly further research is needed, especially with respect to the regulation of human ES cell proliferation.

3.2.3. *Genomic analyses of self-renewal and pluripotentiality*

One approach to study pluripotentiality and self-renewal is through use of genomic techniques. The aim of such studies is to determine the gene expression profiles of undifferentiated cells and determine the gene products that show altered expression once differentiation begins. Molecular methods to analyze mRNA transcript expression include sequencing of Expressed Sequence Tags (ESTs), subtractive hybridization, differential display, competitive PCR, cDNA or oligonucleotide microarrays (Chips), and Serial Analysis of Gene Expression (SAGE) [70]. Because only cDNA microarrays or SAGE permit high-throughput transcriptome profiling, application of these techniques may best provide critical insight into self-renewal and the early steps of specification and differentiation. The assumption in performing genomic experiments such as these is, of course, that self-renewal and pluripotentiality involve a degree of regulation (e.g., transcriptional) that affects mRNA abundance, a logical assumption based on known developmental paradigms (see above).

We have recently performed genomic experiments to determine the transcriptome of pluripotentiality embryonic cells. We hypothesized that identification of factors responsible for establishment and maintenance of pluripotential would permit elucidation of pathways that might promote committed cells to become pluripotent – a critical step in transforming differentiated autologous cells into cells suitable for cell transplantation. Embryonic carcinoma (EC) cells are similar to ES cells, in that they are pluripotent and capable of generating many cell types, after initiation to differentiate by chemical inducers [71-73]. EC cells, however, have a transformed phenotype, are generally derived from malignant teratocarcinomas, and rarely form functional gametes. EC cells are also readily cultured *in vitro* in the absence of feeder layers.

For our transcriptome analyses, we chose to use Serial Analysis of Gene Expression (SAGE). It is the only technique that currently promises a quantitative and qualitative characterization of a cell's complete transcriptome, including mRNAs that have not yet been identified [74, 75]. Two major principles underlie SAGE analyses. First, short DNA sequences are sufficient to identify individual gene products, and second, concatenation (linking together) of short DNA sequences or tags increases the efficiency of identifying expressed mRNAs in a sequence-based assay (Figure 6). The transcript profile generated by SAGE thus relies on short nucleotide sequences (10-14 bp tags) for gene identification. Theoretically, a 14-base tag composed of the four nucleotides (A,C,G or T) present

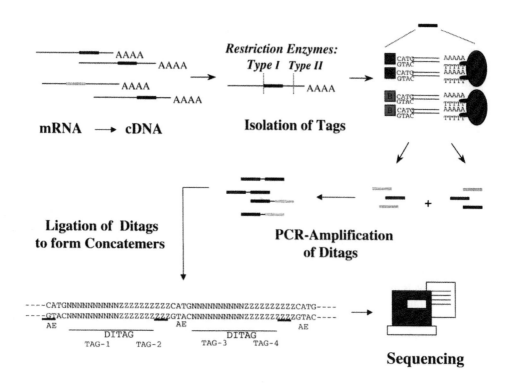

Figure 6. Principal steps of Serial Analysis of Gene Expression (SAGE) (adapted from [74]). Two assumptions are critical for SAGE analyses. One, short DNA sequences (10-14 bp) are sufficient to identify individual gene products, and two, concatenation (linking together) of short DNA sequences or tags increases the efficiency of identifying expressed mRNAs in a sequence-based assay. To generate the sequences, purified mRNA from pluripotent embryonic cells (or any other cell line) is used to generate double-stranded cDNA. Using streptavidin-coated magnetic beads, the double-stranded cDNA is purified, followed by digestion with a type I restriction enzyme or anchoring enzyme (AE) that recognizes specific sites located in the double stranded DNA recognition sequence (CATG for NlaIII). The fragment located closest to the biotinylated primer is then purified by binding to magnetic beads. This fraction is divided in half and ligated to two different linker/primer sets to ensure an accurate quantitative representation of the original transcripts in the sample. SAGE tags are generated by digestion of the cDNA molecules with a type II restriction enzyme or tagging enzyme (TE), which cleaves DNA several base pairs away from the recognition site. The SAGE tags are joined to form ditags and amplified by PCR with a set of primers that recognize Linkers A and B. The ditags are separated from the linkers and ligated together to form concatemers of purified ditags. These are then subcloned into a plasmid vector, amplified and sequenced. The individual tags can then be extracted by identifying the CATG anchoring enzyme sequences. Each individual tag sequence is then run against GenBank databases to identify the corresponding gene product.

in DNA can define 4^{14} (268, 435, 456) different transcripts, i.e., much greater than the 70, 000 to 140, 000 predicted human genes, without any prior knowledge of the gene transcripts present in a cell line.

A SAGE library was generated from undifferentiated mouse P19 embryonic carcinoma cells grown in the absence of feeder layers – a potential contaminant for genomic analyses. Excluding duplicate dimers, >44, 000 SAGE tags were sequenced from this library. Of these, some were excluded from the analysis because they originated from linker sequences, or were determined to be a non-informative, possibly having originated from *Nla*III recognition sites located immediately upstream of a poly (A+) tail. The remaining tags were analyzed further. Two web-linked sites were accessed to identify the mRNAs corresponding to the sequenced tag. The Non-Redundant (NR) Rodent Database and the EST databases were downloaded from GenBank and analyzed using SAGE software. As illustrated in Table 2, not all of the SAGE tags matched gene products in the NR database or entries in the EST database (dbEST).

Of the >43, 000 tags identified in this project, approximately 16, 420 corresponded to unique transcripts. Only 1, 790 tags (10.4%) matched sequences in the NR database, while 10, 264 tags (59.4%) matched sequences in the GenBank EST database. A number of gene products, like Oct3/4, Cripto-1 (see below), pim-2 (Protein-serine/threonine kinase), and msg-1 (melanocyte-specific gene-1), were of particular interest as they had high levels of expression that were later found repressed during differentiation (Table 2). Equally important, factors like osteoblast specific transcription factor-1, H19 (an abundant fetal transcript), and stromal cell derived protein-1, were not present in the undifferentiated cells, but were significantly upregulated during the formation of embryoid bodies (not shown) [53].

Table 2. Example of SAGE tags with expression restricted primarily to P19 EC cells in the pluripotent, undifferentiated state. The examples are limited to gene products whose abundance was rapidly lost following in vitro differentiation of P19 cells to cardiac myocytes. The tag sequences were used to identify matches with GenBank, and the number of tags counted from each unique sequence were obtained from a total of 43,714 sequence tags [53].

SAGE Tag	# of Tags	GenBank Match[1]
1. CATGTGTGGAAACC	43	Protein-serine/threonine kinase (pim-2)
2. CATGGTGGCTCACC	40	Type 2 Alu sequence (M2)
3. CATGGCATAAGGCA	38	EST, similar to Etn transposon
4. CATGCAAAACTGGG	24	Melanocyte-specific gene 1 (msg- 1)
5. CATGGCGGGCGTGG	18	EST, sim. to initiation factor 2-associated 67 kDa protein
6. CATGGTGAAGCTCG	15	EST, similar to Human nucleolin
7. CATGAATATGCACA	10	Tdgf 1 (Cripto- 1)
8. CATGTCCACCCAAA	18	No matches
9. CATGCATTCAAACT	10	Octamer binding transcription factor (Oct-3 and Oct-4)
10. CATGAACTCCACCC	8	EST, no similarity
11. CATGCCTACAAGCC	5	EST, similar to hypothetical 64.0 kD protein

[1] GenBank NR and EST Databases (NCBI)

We have recently completed a SAGE library from R1 ES cells cultured in the absence of feeder layers, but in the presence of BRL conditioned media [31, 60] supplemented with LIF. Although incomplete, we have now sequenced over 100, 000 SAGE tags from R1 ES cells. Similar to the SAGE P19 cell library, >30-50% of the tags did not correspond to sequences in GenBank, and a much larger percentage did not match genes of known function. Importantly, we have performed a preliminary evaluation comparing transcripts from pluripotent EC and ES cells. Analysis of the two libraries has yielded over 150, 000 SAGE tags, corresponding to well over 20, 000 unique transcripts. Of these transcripts, fewer than 200 showed significant differences in abundance. Pluripotent embryonic cells thus have many genes whose pattern of expression is conserved. Of particular interest was the gene *msg-1*. This gene is predominantly expressed in nascent mesoderm, the heart tube, limb bud and sclerotome, and its expression, is restricted, within developing mesodermal sites, to posterior domains [76]. Transcripts for Msg-1 were poorly abundant in undifferentiated R1 ES cells grown in the absence of feeder layers, but they were very abundant in undifferentiated P19 EC cells. These data suggest that a finite number of differences in gene expression exist between pluripotent cells. Msg-1 and a number of other gene products may be involved in cellular transformation (P19 EC cells are derived from embryo-derived teratocarcinomas), but may also be indicative of different cultivation procedures.

We have also used SAGE to identify factors responsible for the early initiating events of embryonic cells induced to differentiate to cardiac myocytes. Numerous known factors implicated in cardiac development (e.g., GATA-4, TEF-1, TGF-β, Hox-7, HAND1) were identified from these analyses, indicating that time-points examined were relevant for the identification of early cardiac regulatory factors. Many novel transcripts, derived from tags corresponding to unknown ESTs or completely unknown mouse sequences, including several with temporally-restricted profiles, show abundant cardiac expression. Three have thus far been identified that have cardiac-restricted expression profiles, while several others have neuronal-expression patterns. These results from SAGE represent the first molecular profile of undifferentiated and differentiated embryonic cells and will be used to identify novel gene products potentially implicated in pre- and early cardiac development [53].

In another set of experiments performed in collaboration with Drs. S. Jaradat and M. Ko (Laboratory of Genetics, NIA), microarray analyses utilizing cDNAs isolated from developmental libraries [52] have been used to address commitment and early differentiation. The aim was to study genetic reprogramming of mouse ES cells during differentiation after withdrawal of LIF, the paracrine factor responsible, at least partially, for directing self-renewal [77]. For this, undifferentiated mouse R1 ES cells were cultivated in the presence of BRL conditioned media and supplemented with LIF. The conditioned media and LIF were withdrawn, total RNAs extracted, and cDNA microarray hybridizations performed. Genes such as STAT3, Oct-4 and Rex1 were apparently down-regulated by withdrawal of LIF, while over 60 other gene products exhibited an expression pattern profile similar to these three. Multiple genes involved in either chromatin remodeling or cell cycle regulation and apoptosis were also clearly responsive to cytokine withdrawal. Over 500 newly identified gene products were affected by LIF withdrawal, the majority of which showed changes in abundance that occurred within a few hours. Based on these preliminary data, we suggest that LIF withdrawal involves the maintenance and repression of hundreds of genes belonging to many gene families, including but not limited to transcriptional activators and repressors, ligand receptors, cell cycle regulators, and growth factors. These data further indicate that loss of pluripotentiality and self-renewal is biologically and molecularly very complex. Functional analyses of individual gene products, identified either by SAGE or microarrays, must however be completed before we will be able to determine which factors or signals are involved specifically in self-renewal, pluripotentiality or differentiation of cardiac myocytes. Analyses of these gene products may also provide critical insights into biological signals regulating cell cycle checkpoints or regulatory factors that specify the cardiac phenotype. Such findings would be a major step toward reinitiating cell growth in post-mitotic cardiac cells or in converting non-cardiac cells into cardiac myocytes.

3.2.4. Gene-trapping and identification of cardiac developmental genes

Identification of developmental cardiac genes, through gene trapping or other techniques, offers important insights into congenital heart diseases [78]. Congenital heart disease occurs in about 5-10 births per thousand. The disease is not always life-threatening and can be managed either through minor surgery or non-surgical means; however, in about 30% of the cases, the lesions require immediate attention [79]. The mechanisms underlying congenital cardiac malformations are frequently not known, and the development of cell replacement therapies based upon the genetic origin of a disease will ultimately depend on an understanding of the genetic and developmental basis of the disease.

The random insertion of exogenous DNA into single sites in the mammalian genome (gene trapping) provides a genome-wide strategy for functional genomics. When performed in tandem with the developmental potential of ES cells to differentiate into distinct cell lineages, such techniques are useful for the identification of novel genes expressed in developing systems (see Figure 4) [80-84]. Gene trapping is thus a powerful experimental approach to identify developmentally regulated genes in mouse embryos or ES cells. Effectively gene trap constructs, consisting of a resistance gene (e.g., neo[R]) and a promoterless LacZ reporter gene with a splice acceptor consensus sequence at its 5' end, are introduced into cultured ES cell lines to disrupt normal gene function. Resistant ES cell colonies are selected and expanded as undifferentiated cells *in vitro*. These cells can then be either injected into blastocysts to generate chimeric mice or differentiated *in vitro* (see below). Expression of the gene trap is assayed for by β-galactosidase (β-gal) staining and is indicative of an insertion event within a transcriptional unit. *In vivo* gene trap screens, in mice, permit identification of genes that are expressed either within specific tissues or in spatiotemporal patterns. Alternatively, gene trap expression within specific cell types can be assayed in differentiated ES cell cultures. These cultures provide a simple model system for studying the genetic pathways that regulate embryonic tissue development. Such a technique, particularly when used in conjunction with *in vitro* pre-selection [85, 86], permits high-throughput screening of clones for tissue-restricted gene trap expression, while eliminating those clonal lines with mutational events in ubiquitously expressed house-keeping genes.

This technique is rapid and precludes generation and examination of hundreds of chimeric embryo litters. Forrester et al. and Stanford et al. successfully used this system to identify several novel retinoic acid responsive genes and hematopoietic and vascular genes, respectively [85, 87].

Baker et al. used *in vitro* pre-selection to study the molecular mechanisms regulating specification and commitment of a variety of cell and tissue types, including heart [86]. Using gene traps, the authors were able to identify single copy genes that were expressed in the developing nervous system, heart and limb cartilage. One of the *in vitro* differentiated genes expressed in cardiac myocytes and identified by gene trapping corresponded to *jumonji* (*jmj*) [86]. The *jmj* gene had previously been identified as developmentally important in liver, spleen, thymus and nervous system, but its role in developing heart was not understood [88]. The *jmj* homozygous mouse embryos were found to have heart malformations, including ventricular septal defects, non-compaction of the ventricular wall, a double-outlet right ventricle and dilated atria [89, 90]. The JMJ protein was, furthermore, found to be a nuclear factor that played a critical role in normal heart development [89, 91]. Examples such as this indicate how gene-trapping techniques, in conjunction with *in vitro* and *in vivo* assays, can be used to identify important developmental genes associated with cardiac differentiation. At this time, it is entirely unclear if JMJ plays a role in aging or heart disease. Findings, such as those for JMJ discovered through the use of ES cells, are critical to determine the molecular basis of defects associated with congenital heart disease, and in some cases heart failure.

3.3.   ES cell-derived cardiac myocytes

3.3.1.   *Properties of ES cell-derived cardiac myocytes*

Embryonic stem cells readily differentiate to cardiac myocytes when cultured under appropriate conditions. A number of parameters influence the developmental potency of ES cells in culture: 1) the number of cells differentiating in the embryoid bodies; 2) media, quality of fetal calf serum (FCS); growth factors and additives; 3) ES cell lines; and 4) the time of embryoid body plating [92]. For the development of ES, EC or EG cells into differentiated phenotypes of cardiogenic muscle cells, pluripotent cells can be cultivated as embryoid bodies (EB) by either the 'hanging drop' method [93], in mass culture [25], or by differentiation in methylcellulose [94]. When appropriately followed, pluripotent mouse embryonic cells readily produce cardiac myocytes *in vitro*.

We use a differentiation protocol employing the 'hanging drop' method (Figure 7). ES cell suspensions containing a defined ES cell number (400, 600 or 800 cells) are prepared in differentiation medium [92, 102]. Drops of cell suspensions are placed onto the lids of petri dishes and cultivated as hanging drops for 2 days. These cells spontaneously aggregate to form embryoid bodies. The EBs are transferred to suspension cultures for 3-5 days, before plating in microwell plates for morphological analysis, or transferred to tissue culture dishes. Within 1-4 days, spontaneously contracting cardiac myocytes can usually be seen.

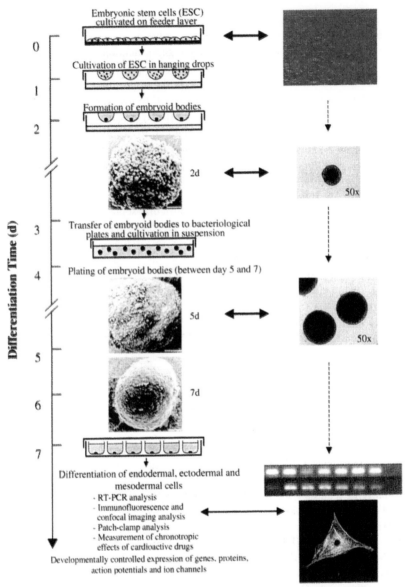

Figure 7.   Differentiation protocol for embryonic stem (ES) cell derived cardiogenic differentation (adapted from [42, 165]). Undifferentiated ES cells cultivated on primary cultures of embryonic fibroblasts (feeder layers) are cultivated as embryoid bodies (EBs) in hanging drops for 2 days and in suspension for 3 to 5 days before plating onto gelatin-coated tissue culture dishes. The morphology of ES cells on feeder layers, 2d-, and 5-day-old-EBs are shown by light microscopy on the right; while the morphology of 2-, 5-, and 7-day-old EBs by scanning electron microscopy (bar = 50 μm) is shown on the left. After differentiation, numerous properties of the ES cell-derived cardiogenic cells can be studied. A number of the potential measurements are listed. Examples of RT-PCR analysis of β-tubulin and RyR2 (time points examined from left to right: day 5, 7, +1, +3, +5, +7, +14 and water control) and an immunoflourescence image of a cardiac myocyte stained with an antibody to troponin T are shown.

Several developmental stages in the differentiation of cardiac myocytes have been described [93, 95-102], and a number of cardiac genes are activated during the *in vitro* differentiation of ES cells (Table 3) [103]. ES cell-derived cardiac myocytes, for example, express a number of transcription factors known to be important in early cardiac development, including GATA4, dHAND, eHAND and Nkx2.5 [42, 104]. Markers characteristic of the cardiac phenotype are also expressed, including α- and β-myosin heavy chain, α-tropomyosin, myosin light chain-2v, atrial natriuretic factor, phospholamban, L- and T-type calcium channels (unpublished data), SR CaATPase and type 2 ryanodine receptors (RyR2). Sarcomeric proteins of ES cell-derived cardiac myocytes were established in a developmentally controlled manner in the following order: Titan (Z-disk), α-actinin, myomesin, titin (M-band), myosin heavy chain, α-actin, cardiac troponin T and M protein [105]. This pattern was found to be in agreement with those from skeletal muscle development [106] and chicken heart development *in vivo* [107, 108]. Developmental transitions in isoforms of troponin I and MHC expression have also been detected in cultures of differentiating ES cells [103, 109]. Furthermore, transfection studies with cardiac-restricted promoters have shown restricted patterns of gene expression in ES cell-derived cardiac myocytes [110-112]. These patterns of gene expression and others, therefore, mimic many of the known changes in expression seen during early cardiac development *in vivo*.

Spontaneously and rhythmically contracting cardiac myocytes can be isolated by mechanical dissection and collagenase treatment. Individual cells can then be used in single cell assays (see Figure 7). Depending on the time at isolation, the ES cell-derived cardiac myocytes have electrophysiological characteristics typical of early pacemaker like cells, but with further differentiation, have action potentials typical for atrial, ventricular and conduction system cardiac myocytes (Table 3). Terminally differentiated ES cells also show pharmacological and physiological properties of specialized myocardial cells, typical of atrial-, ventricular-, purkinje- and pacemaker-like cells [97, 98, 102, 113]. Functionally, these cells show normal contractile sensitivity to calcium, and exhibit many of the properties associated with normal excitation-contraction (EC) coupling [114]. The cells *in vitro* do not, however, seem to form T tubules or show some morphological characteristics typical of adult cardiac myocytes.

Cardiac myogenesis thus is recapitulated during differentiation of ES cells to cardiac myocytes. The rapid *in vitro* development and ease of genetic manipulation of ES cell culture systems, together with contractile and electrophysiological assays, provide investigators with a valuable tool to study cardiac development [97-99, 103]. Specifically the system can be used to determine cardiac gene structure-function relationships *in vitro*, test potential cardioactive drugs, and examine regulatory mechanisms of cardiac-restricted gene expression. Finally, the use of ES cell cardiac myocytes as a potential source to test the feasibility of using donor cells for cardiac cell therapy represents an exciting application of this system for clinical purposes.

3.3.2. *Functional aspects of in vitro ES cell-derived cardiac myocytes*

Intracellular signaling by calcium ions is a ubiquitous mechanism in mammalian cells. Up to 80% of all signaling in mammalian cells involves the movements of Ca^{2+};

Table 3. Summary of cardiac-specific properties of *in vitro* differentiated embryonic stem (ES) cells (adapted from [42] and unpublished). The mRNA abundance, protein abundance, action potential characteristics and ion currents from early stages of differentiation until terminal differentiation are shown. Embryoid bodies were cultivated as described in Figure 7.

A.

Stages:	Early		Intermediate								Terminal
	3	5	7	+2	+4	+6	+8	+10	+12	+14	+18

Time of Differentiation (d)

B. mRNA Abundance

	3	5	7	+2	+4	+6	+8	+10	+12	+14	+18
Nkx2.5	±		++			++				+	
α,CaCh	+	+	++			++	++	++		++	++
α-MHC		+	++		++	++	++	++	++	++	++
β-MHC		+	++		++	++	++	++	++	++	++
ANF			±		+					++	
MLC2v			+	++	++	++	++	++	++	++	
SERCA2	±		++	++	++	++	++	++	++	++	
RyR2	±		+	+	+	+	+	+	+	+	
NCX1	±		++	++	++	++	++	++	++	++	

C. Protein Abundance

	Early	Intermediate	Terminal
Titin (Z-band)		++	++
α-actinin		++	++
Myomesin		++	++
Sarcomeric MHC		+	++
Sarcomeric α-actin		+	++
cardiac Troponin T	-	±	+
M-protein	-		+
RyR2	± +	+	+

D. Action Potentials

Early   Intermediate   Terminal stage (Atrial-like, Ventricle-like, Sinusnodal-like)

E. Ion Currents

	Early	Intermediate	Atrial-like	Ventricle-like	Sinusnodal-like
I_{Ca}	+	++	++	++	++
I_{to}	+	+	+	+	+
$I_{K,ATP}$	+	+	+	+	+
I_{K}	-	+	+	+	+
I_{Na}	-	+	++	++	±
I_{K1}	-	-	+	+	-
$I_{K,Ach}$	-	-	+	-	+
I_{f}	-	+	-	-	++
$I_{Ca,Ryr}$	±	+	+	++	+

Table 3. **Legend to Table 3 - continued. A.** Stages of ES cell-derived cardiac myocyte differentiation and the time points studied. Almost all of the EB outgrowths, between 2 and 10 days after plating, contained areas with spontaneously beating cardiac myocytes. After this time, the number of EBs with spontaneous contractions decreased. **B.** The mRNA abundance of selected cardiac genes was studied by RT-PCR at selected time points during early, intermediate (7 +2-+4days), and terminal stages of differentiation (, very low expression, +, low expression, ++, strong expression). **C.** The accumulation and structural organization of sarcomeric proteins in isolated cardiac myocytes was determined by immunofluorescence (Titin, α-actinin, myomesin, myosin heavy chain (MHC), α-actin, cardiac troponin T and M-protein) or by Western analysis (type 2 ryanodine receptor – RyR2). **D. and E.** Electrophysiological analyses revealed functional properties of early and terminally differentiated ES cell-derived cardiac myocytes. Cells expressing pacemaker-like action potentials were only found at early and intermediate stages, while atrial-, ventricular- and sinusnodal-like types of action potentions were identified primarily in terminally differentiated cells. (I_{Ca}, L-type Ca^{2+} channel; I_{to}, transient K^+ channel; $I_{K,ATP}$, ATP-modulated K^+ channels; I_K, delayed outwardly rectifying K^+ channels; I_{Na}, Na^+ channels; I_{K1}, inwardly rectifying K^+ channels; $I_{K,Ach}$ muscarinic acetylcholine-activated K^+ channel; I_f, hyperpolarization-activated pacemaker channels; $I_{Ca,Ryr}$, sarcoplasmic reticulum ryanodine-sensitive calcium release channel)

---

however, this signaling depends on the spatial and temporal distribution of calcium ions. Movements of calcium in the same cell can thus convey more than one signal and response. This is particularly important in neurons, where the signals responsible for the long-term synaptic changes involved in memory are conveyed, in part, by intracellular [Ca^{2+}], and in the heart, where localization of calcium signals is crucial to the regulation of excitation-contraction coupling (ECC). The role of ECC in ES cell-derived cardiac myocyte differentiation has been poorly characterized.

The contraction of cardiac muscle, like many other cellular processes, is controlled by the level of cytosolic free calcium, [Ca^{2+}]$_{cyto}$. The contraction of myofilament proteins is triggered by a transient rise in [Ca^{2+}]$_{cyto}$. The coupling of electrical excitation of the heart to the production of contraction (ECC) involves the interaction of a number of cellular proteins involved in Ca^{2+} homeostasis. The calcium transient is signaled by the entry of Ca^{2+} through voltage-controlled L-type calcium channels in the sarcolemma, but, in mammals, most of the [Ca^{2+}]$_{cyto}$ comes from stores in the sarcoplasmic reticulum (SR). SR calcium is released in response to the entering calcium by a process of *calcium induced calcium release* (CICR), mediated by type 2 ryanodine receptor release channels (RyR2). Together both sources of Ca^{2+} initiate contraction. Relaxation is brought about as [Ca^{2+}]$_{cyto}$ is pumped into the SR by the SR phospholamban (PLB)-regulated Ca^{2+}-ATPase (SERCA) and extruded from the cell primarily by the Na^+/Ca^{2+} exchanger (NCX1). Thus in the adult myocardium, Ca^{2+} from both the intracellular and extracellular sources are required to induce muscle contraction [50].

In the developing myocardium, it is generally believed that primarily extracellular and not intracellular stores of Ca^{2+} are critical for contraction of the cardiac myocytes.

The developing myocardium in rodents has a distinct temporal and spatial molecular gene expression profile that has been postulated to determine, at least in part, its morphological and functional characteristics [115, 116]. Graded expression of SERCA2 and PLB and homogeneous expression of RyR2 and NCX1 transcripts and protein permit the compartments of the developing myocardium to contract and relax in a coordinated manner [115-118]. Blood thus moves unidirectionally in the embryonic and fetal myocardium even in the absence of valves. These data suggest that regulated control of internal stores of Ca^{2+} is essential to normal myocardial ECC. Attempts to address the structural components of this system have however been difficult, because the inactivation of SERCA2 and type 2 RyR gene expression in knock-out mice results in embryonic lethality [119, 120]. Although embryonic lethality in these mice suggest that regulation of $[Ca^{2+}]_{cyto}$ is critical to normal development of the myocardium, it is impossible to determine what functional role they play in differentiation or in regulating Ca^{2+} homeostasis with development. Equally difficult to study are other genes which are essential for early developmental events.

3.3.3.   *Loss-of-function studies* in vitro

Differentiation of ES cells and formation of embryoid bodies have provided a major opportunity to study the loss-of-function of genes involved in early stages of differentiation or which result in embryonic lethality [42]. Because ES cells are diploid, production of ES cell knock-outs is generally more laborious than that necessary for creation of targeted mice. Use of this model system requires targeting of alleles on both chromosomes (see Figure 4). Once achieved, knockout of specific genes can be used to test the ES cell's ability to differentiate to cardiac myocytes, or any other cell type where defined protocols are available. Special care should be taken to ensure that non-targeted alleles are not mutant (random insertion events, translocation, altered karyotypes). When controlled for, however, these cells represent a unique system to study genes that otherwise would be early embryonic lethal.

   Recent examples of ES cells lacking specific functional genes include deletion of GATA-4 [104], Desmin [44], Cripto-1 [121] and β1 integrin [43]. Cripto-1, for example, is a growth factor with an epidermal growth factor-like motif encoded by the *teratocarcinoma-derived growth factor-1* gene. Inactivation of *cripto-1* results in the lost ability to form beating cardiac myocytes but not in the differentiation to other cell types like skeletal muscle. Cardiac-specific gene transcripts are also absent. These findings suggest a potential role for Cripto-1 in the process of gastrulation and the regulation of cardiac gene expression. β1-integrin, in contrast, is a member of a large family of cell surface receptors that bind to extracellular matrix proteins and cell counter receptors. Inactivation of this gene is lethal to a developing mouse fetus (see Figure 4). *In vitro*, ES cells lacking the β1 integrin result in a delayed expression of cardiac muscle-specific genes and action potentials, and a reduction in the occurrence of atrial- and ventricular-like cells. Cardiac myocyte sarcomeric architecture is also greatly disturbed by loss of this gene. β1 integrin thus seems to be required for differentiation and maintenance of cardiac myocyte cell phenotypes. These data on gene function were only possible because of the generation of knock-out alleles in ES cells.

To address the role of RyR2 in ECC with development, we have prepared several neomycin-resistant R1 ES cell clones lacking an essential exon for this gene (*ryr2*) [122]. Knock-out of *ryr2* results in embryonic lethality [120], but in the context of this recombination event, cardiac myocytes differentiate similarly to control cells – all of the embryoid bodies give rise to spontaneously contracting cardiac myocytes. The *ryr2* knock-out cells, like control cells, have spontaneous and triggered increases in cytosolic Ca^{2+}. The knock-out cells lack Ca^{2+} transients evoked by caffeine, an activator of RyR Ca^{2+} release. Gene expression profiles for cardiac gene products were similar to those from control cells. These data from the *in vitro* differentiated *ryr2* knock-out ES cells indicate that Ca^{2+} release from the SR is not required for cardiac differentiation. A more likely explanation for the lethality in *ryr2* knock-out mice probably relates to the channels role in regulating cardiac function and blood flow in the developing embryo.

3.3.4. *Lineage selection of ES cell-derived cardiac myocytes*

In vitro differentiation of pluripotent ES cells leads to the formation of many types of cells that grow together in culture. One of the major problems for researchers has thus been how to isolate a pure population of specific ES cell-derived differentiated cells potentially useful for cell therapy. Several groups have recently been successful in isolating individual cell lineages or types using selection protocols, genetic manipulation and *in vitro* differentiation of ES cells.

The use of specific growth conditions and genetic approaches has led to rapid progress toward lineage selection. For example, non-mesodermal precursors, like neuronal cells, have been derived from proliferating neuroepithelial precursor cells. This was accomplished by suppressing, almost completely, mesodermal differentiation through growth factor depletion [123]. Use of these strategies was subsequently successful in transplantation experiments using animal models of human disease [124, 125]. Similar strategies are currently being used to isolate dopaminergic and serotonergic neurons [126]. The differentiation and selection protocols involve several steps, including embryoid body formation, growth and selection of central nervous system stem cells, and withdrawal of basic fibroblast growth factor (bFGF) for differentiation to dopaminergic neurons. The authors found that midbrain and hindbrain neurons could be generated in unlimited numbers from ES cells, and they suggest that genetic manipulations could now be used to obtain pure populations of cells suitable for grafting [126]. Li et al. described a system which allowed for the efficient purification of neuroepithelial progenitor cells [127]. Specifically, they targeted a promotorless neomycin cassette to the SOX2 gene, so that only those cells expressing SOX2 following *in vitro* differentiation would be neomycin (G418)-resistant. The SOX2 gene encodes a DNA binding protein whose expression is restricted to the early neural plate. The G418-resistant cells were shown to differentiate into neuronal networks in the absence of other cell types. Dinsmore et al. also reported the differentiation of ES cells into populations of neurons and skeletal muscle cells [128, 129]. For skeletal muscle cell isolation, ES cells were transfected with a gene coding for the muscle-specific regulatory factor MyoD and resistant clones isolated [128]. When these ES cell clones were induced with DMSO, the cells expressing high levels of MyoD differentiated into skeletal myoblasts that fused and formed myotubes capable of contraction.

When optimal conditions are used to induce differentiation to cardiac myocytes, the percentage of differentiated cells belonging to one phenotype remains small (in the case of cardiac differentiation at intermediate to late stages, this percentage is frequently of the order of 3-5%). The group of L. Field has demonstrated that genetically selected cardiac myocytes can be isolated from differentiated ES cells and form stable intracardiac grafts [130]. They prepared an ES cell line carrying a fusion gene comprised of the cardiac α-myosin heavy chain gene promoter and a neo[R] cassette. After selection for ES cell clones that were transfected stably, 99.6% of the G418 resistant cells consisted of cardiac myocytes. Other groups have not been as successful in isolating a pure population of cells using the same promoter. The cardiac myocytes isolated by Field's group were injected into the ventricular myocardium of *mdx* mice and found to be present in the grafts for at least 7 weeks after implantation. These results validate the potential use of ES-derived cells in cardiac cell therapy. At present no long-term animal experiments in heart have been performed to demonstrate that transplantation of ES cell-derived donor cells do not give rise to tumors.

We have used similar techniques with multiple 'cardiac' promoters to isolate cell specific lineages with selection cassettes, including neo[R] and puromycin resistance (pur[R]) cassettes. The results have not produced a 'pure' population of cardiac myocytes, but have yielded important insights into lineage selections. To isolate a pure population of ventricular myocytes and examine their developmental potential, we generated stable ES cell lines containing a puromycin-resistance cassette driven by a CMV-enhancer and a 2.1 kb myosin light chain 2v promoter. Normal R1 ES cell clones positive for *pur[R]* were differentiated *in vitro* and examined. Differentiated cardiac cells from the puromycin resistant clones displayed spontaneous beating profiles typical of control ES cell populations. Transfected, but non-PUR-selected and normal R1 ES cell-derived cardiac myocytes contain mRNA for RyR2, SERCA2, MLC2v, β-MHC and α-MHC. In PUR-resistant cells, transcripts for RyR2, SERCA2 and markers of ventricular cells MLC2v and β-MHC are present, while markers for atrial cells, α-MHC and MLC2a, are absent. Indirect-immunofluorescence indicated that most of the cells that survived puromycin selection were muscle cells (>90%). Electrophysiologically, the cells had ventricular characteristics and no response to carbachol, a functional marker of SA nodal and atrial cells. In older cells, action potentials were ventricular-like in origin. The data suggest that the majority of the puromycin-resistant cells are ventricular in origin. We also found that the dose of puromycin used in the selection process was critical for isolation of cardiac myocytes. While high doses of puromycin, sufficient to kill >95% of non-resistant cells in less than 48 hours, gave acceptable numbers of cardiac myocytes, the cells did not display good electrophysiological characteristics. Lower doses of puromycin with longer selection periods improved the electrophysiological characteristics, but did not select for comparable pure populations (unpublished data).

ES-derived differentiated cells can also be isolated without antibiotic selection through use of a fluorophore expressed in specific types of differentiated cells. Metzger et al. first engineered a D3 ES cell line to regulate the LacZ gene with the cardiac α-actin gene promoter [109]. Kolossov et al., used the same promoter to select for ES differentiated cells expressing the green fluorescent protein (EGFP) [111]. At early steps in the differentiation protocol, only cardiac myocytes expressed the green fluorescent protein, and they could be isolated by FACS methods. At later times of differentiation, non-cardiac myocytes also expressed the green fluorescent protein, limiting its usefulness to primarily early cardiac development. More recently, the MLC2v promoter has been used to direct enhanced cyan fluorescent protein (ECFP) and EGFP expression in ES-differentiated cells [112, 131]. Although the promoter construct used in the experiments by Meyer et al. is not necessarily restricted to ventricular cells (expression of this promoter element is in ventricles and slow-skeletal muscle), ECFP was expressed exclusively in cardiac ventricular cells. These cell clones, thus, may represent a model system to study signals regulating cell proliferation and terminal differentiation of cardiac cells destined to become ventricular.

All these data show that specific and relatively pure cell types can be isolated from *in vitro* differentiated ES cells; however, before these cells can be useful in cell therapy protocols, better techniques will be required to preserve the physiological characteristics of the selected cells. Whether similar success will be forthcoming from human ES cells is unknown, but the prospects seem favorable.

3.4.   Transgenic mice and creation of new animal models

Mammalian embryos are extremely resilient in the early stages of development and can tolerate both the loss of tissue, and incorporation of cells from other embryos. This property has enabled the incorporation of foreign cells into the developing embryo early in development, permitting the elucidation of cell lineage and the investigation of cell potential. The existence of ES cells has also allowed the creation of a wide variety of mutations in any given gene through the use of techniques like homologous recombination or gene trapping which can be incorporated into the mouse genome (see Figure 4). The extent of contribution of foreign cells depends on the cell's normality, genotype, mitotic and developmental potentials, and developmental synchrony with the host embryo. If transmission to the offspring of the altered genotype is desired, then foreign cells must have the capacity to undergo meiosis and gametogenesis. Generation of chimeras of genetically defined offspring from targeted ES cells has been extensively reviewed in the literature (see [37, 132-141] for further information). So far, most animal models have been generated to mimic human pathologies and success has almost exclusively been achieved with mouse ES cells transferred to generate mice.

4.   Pluripotent human embryonic cells: ES and EG cells

4.1.   Overview

The derivation of human ES cell lines derived from human embryos or fetuses represents the first major step in obtaining a renewable, tissue culture source of human cells capable of generating into a wide variety of cell types, including cardiac myocytes.

The implications that relate to the development of these cells have broad applications for basic research and potential clinical use in humans. The question is whether the human ES cells will be as versatile as the mouse ES cells with respect to self-renewal and developmental capacity, both of which are required for their use in basic research and clinical applications.

In 1998, Thomson et al. reported the derivation of the first ES cell lines from human blastocysts obtained from *in vitro* fertilized human eggs [20]. Several characteristics of ES cells were determined including a preference for growth on feeder layers with LIF, a high nucleus to cytoplasm ratio, prominent nucleoli, and colony morphology similar to that of ES cells derived from rhesus monkey. One of the ES cell lines isolated retained a normal XX karyotype even after 6 months of culture. This human ES cell line expressed high levels of telomerase activity and displayed cell surface markers characteristic of undifferentiated nonhuman primate ES and human EC cells. The undifferentiated cells showed positive alkaline phosphatase activity. Upon injecton into severe combined immunodeficient (SCID) beige mice, teratomas formed and derivatives of all three germ layers were identified. With these tests, the authors presented evidence that human ES cells have in fact been derived.

Shamblott et al. subsequently reported the derivation of pluripotent stem cells from cultured human primordial germ cells [21]. These human EG cells showed many of the characteristics of pluripotentiality as described for mouse ES/EG cells: high levels of alkaline phosphatase activity, compacted multicellular colonies, positive staining for a number of appropriate germ-line markers, and maintenance of normal karyotypic features. In a recent paper, the authors determined that the original EG cells had a limited proliferation capacity, and only embryoid body-derived (EBD) cells proliferated and differentiated into cells expressing neuronal, skeletal muscle, vascular, hematopoietic and endodermal markers [142]. Although a limited number of EBD cells were examined, the differentiation capacity of these cells did not show properties typical of cardiac myocytes, restricting the potential use of these specific cells in cardiac cell therapy protocols.

4.2.   Properties of pluripotent human ES cells

Human ES cells have a number of characteristics that are similar to, but distinguishable from mouse ES cells (see Table 1). As described previously, ES cells require an embryonic fibroblast feeder-cell layer for culture, but in contrast to mouse, feeder layers can not be replaced by LIF alone when cultivating human ES cells. Mouse ES cells grow as colonies with a smooth, rounded morphology; whereas, human ES cells grow in flat colonies with distinct cell borders. All pluripotent human and mouse stem cell lines have alkaline phosphatase activity. A set of cell surface antigens, however, distinguishes ES cells derived from the two sources. Human ES cells express SSEA-3, and -4 in the undifferentiated state, and SSEA-1 only upon differentiation [20]. Undifferentiated mouse ES cells express SSEA-1, but not SSEA-3 or –4 [143]. Both mouse and human ES cells express Oct-4, but beyond this finding, very little is known about conserved patterns of gene expression between these two sets of pluripotent stem cells. A number of conserved molecular markers must be present, but a thorough analysis, preferably by SAGE, is necessary before any conclusions can be drawn. The work described earlier comparing mouse ES and EC cells by SAGE analysis does, however, bode well for determining what factors contribute to self-renewal and pluripotentiality in human ES cells.

4.3. Use in basic research and toxicology

The acquisition of human ES cells has important implications for basic science investigations. Human ES cells can be used to study, *in vitro*, normal human embryogenesis, abnormal development (generation of cell lines with modified genes and chromosomes), functional genomics (human gene discovery), pharmacotoxicology (drug screening), and embryotoxicity of potential teratogens (Figures 3 and 4, *in vitro*). Applications of human ES cells, however, are based on the assumption that ES cells can be grown on a large scale, that genetic modifications can be introduced, and that researchers will be able to identify protocols that direct their differentiation to specific cell types *in vitro*. These potential applications have yet to be realized, but are on-going.

The differentiation of human ES cells into embryoid bodies or into teratomas is currently spontaneous and uncontrolled, but a number of important characteristics typical of mouse ES cells are present. Cultures of human ES cells require growth on feeder layers, because LIF alone appeared insufficient to maintain long-term pluripotentiality [144, 145]. A transcription factor associated with mammalian pluripotentiality, Oct-4, is expressed in human stem cells, and its expression is down-regulated during differentiation [22]. Induction of cystic embryoid bodies can lead to regional expression of embryonic markers specific to different cellular lineages that contain cells of neuronal, hematopoietic and cardiac origins [146, 147]. The differentiation process of human ES cells *in vitro* can be modulated by growth factors. Activin-A and TGF-β1 mainly induced mesodermal cells, retinoic acid, EGF, BMP-4 and bFGF activated ectodermal and mesodermal markers, while the growth factors NGF and HGF allowed differentiation into endoderm, ectoderm and mesoderm. Cardiac myocytes are prominent in cultures supplemented with HGF, EGF, bFGF, and retinoic acid. None of these factors, however direct ES cell differentiation exclusively to a single cell type [146]. These data indicate that human ES cells have, at least some, important characteristics in common with mouse ES cells, including a requirement for fibroblast feeder layers, expression of proteins thought to be important for self-renewal, and enrichment of cell types during differentiation with specific growth factors. Whether these cells are amenable to genetic manipulation is still unresolved. The characterization of these attributes, however, represents the first step toward characterizing a renewable source of pluripotent human cells, both for basic research and potentially for use in cell therapies.

4.4. Cell replacement therapies

The proof of cell replacement therapy in heart has already been shown experimentally in mouse, not only with modified mouse ES cells as described, but also with C2C12 myoblast and fetal cardiac myocyte grafts [130, 148, 149]. Cultivation *in vitro* and isolation of specific differentiated cell types thus may ultimately lead to the use of human ES cells as a renewable source for tissue transplantation, cell replacement, and gene therapies. Human ES cell cultures may also preclude the direct use of fetal tissues in transplantation therapies, a potential boon for those who object to the use of fetal tissue for any medical or research purpose. Clinical targets for therapy might include neurodegenerative disorders, diabetes, spinal cord injury, and hematopoietic repopulation and myocyte grafting. Even if stem cells are not available for transplantation therapy in humans, they offer the possibility of creating animal models to study the fate of human cells or to discover drugs that affect self-renewal, pluripotentiality and differentiation. A number of studies have shown that it is possible to graft human cells in SCID mice and, provided that stem cells are available, it might be possible to create an animal in which a tissue (or part of it) is human. Such examples have already been described in the case of bone marrow-derived cells [150]. Animals of this sort should prove instrumental in the analysis of both normal and pathological development. An example of the potential of this approach has been published in the case of the McCune-Albright syndrome [151]. Of course, these experiments have substantial ethical implications that warrant further consideration.

Before transfer of ES-derived cells to humans can proceed, a number of experimental obstacles must be overcome. Will ES-derived cells be autologous/immunologically masked and be expandable *in vitro*? Will selection vectors integrate stably into either targeted or inconsequential random genomic sites without adverse genetic effects, and from these genetically modified cells, can homogeneous, pure cell populations with defined phenotypes be prepared? In the scenario where human ES-derived differentiated cells (cardiac myocytes, neuronal cells, hematopoietic cells or vascular smooth muscle cells) can be isolated, appropriate steps will need to be taken to prevent rejection of the transplanted cells or prevent cell contamination that might lead to the spontaneous formation of teratomas or teratocarcinomas. It will be important to guard against karyotypic changes during passaging and preparation of genetically-modified ES-derived cells. Without such controls, there is no assurance that consequential changes in the genome will not occur.

5.   Adult stem cells *versus* embryonic stem cells

Adult stem cells (ASC) are progenitor cells that usually give rise to subpopulations of specific cell types (see Figure 1), many properties of which have been discussed elsewhere in this volume. Importantly, recent studies have indicated that some adult stem cells can be coaxed into differentiated cells not normally associated with their 'committed' state [152]. Examples include hematopoietic stem cells from bone marrow that were found to develop into neural, myogenic and hepatic cell types, neural or skeletal muscle stem cells that developed into the hematopoietic lineage [152-158], and stromal stem cells differentiating into cardiac myocytes [58, 159]. The question thus arises as to whether ASCs are pluripotent, similar to ES cells. A major advantage in the use of ASCs for cell replacement therapy is that they will not provoke immune-system rejection, should not become malignant, and may differentiate into a finite number of cell types. In transplantation scenarios, prevention of immune-responses is critical to success, because all stromal-derived and hematopoietic-derived cells (except red blood cells) express either HLA class I or class II molecules. When the donor cells are not autologous, the HLA class molecules can be recognized by specific lymphocytes that may destroy cell grafts. ASCs derived from an individual needing cell therapy would not succumb to any immunological response; whereas, ES-derived cells probably would elicit strong immunological responses in the absence of immunosuppressive therapy or allogeneic compatibility.

Based on our present knowledge, adult stem cells, when compared to ES cells, do not have the same developmental capacity. For example, injection of ASCs (hematopoietic or neuronal) into a mouse blastocyst can contribute to a variety of tissues, but the contribution differs in each embryo. ASCs do not contribute to all tissues of the developing embryo [153] and, *in vivo*, seem to have a limited differentiation capacity. Obviously, somatic stem cells of the adult organism may yet have a high plasticity and their developmental potential may not be restricted to one lineage, but could be determined by the tissue environment in the body [160]. The identification of such reprogramming factors will be one of the challenges of the future. These studies will show whether it may be possible to reprogram, not only adult somatic nuclei by fusion to enucleated eggs [161], but also to (retro- and/or trans-) differentiate adult somatic stem cells in response to 'reprogramming' factors.

The use of genomic techniques such as SAGE or microarrays will eventually identify differences between ASC and ES cells in signaling pathways triggered with proliferation and differentiation. At the moment, ASCs do not hold the same promise as mouse ES cells to be a robust and renewable resource for cell therapy. As we learn more about these cells, they may, however, have the potential for differentiation to specific cell types that will be extremely useful, and immunologically preferable to ES cells for cell therapy.

The increased incidence of the large offspring syndrome from the cattle and sheep industry, however, indicates some of the potential dangers of epigenetic phenomena associated with the preparation and growth of cells *in vitro* for transfer to blastocysts [162]. In the production of the cloned sheep Dolly by nuclear transfer technology, it was assumed that her genome would carry the mutations that had accumulated with aging prior to the transfer [163]. Dolly is, however, still alive and her offspring apparently normal. Perhaps this bodes well for the grafting of genetically modified and differentiated cells that have been passaged many times in tissue culture. The potential for genetic mutations in cells to be used in cell transplantation cannot however be ignored. Perhaps such scenarios will not prove of consequence with sufficiently differentiated and selected cells; however, other techniques or rapid screening procedures (e.g., microarray technology) may be necessary to control the quality and purity of the cells (ES cell- or ASC-derived) used for transplantation. Given adequate public support, the impact of human ES and adult somatic stem cells on human biology and medicine could however surpass that brought about by the basic studies of mouse embryonic cell lines.

6.   Conclusion

Cardiovascular diseases are still the leading cause of morbidity and mortality in industrialized nations. In the aged, these diseases reach epidemic proportions and foreshadow a potential worldwide crisis later this century. While modern cardiology has dealt effectively with some aspects of these diseases, the most severe cases of heart failure require cardiac transplantation. Unfortunately, suitable donor material is in extremely short supply. Perhaps therapies, involving cell replacement, may be able to supplement or replace the need for transplantation. Clinical interventions of this type will, however, require a supply of cells that can be grown on a large scale, can be manipulated as a pure cell population, and can be grafted to damaged or diseased tissues (Figure 8). One exciting possibility to accomplish these goals may be application of human ES cells, as well as ASC (wherever possible), to cell replacement therapy.

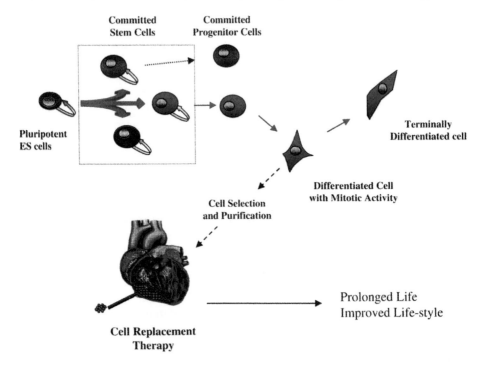

Figure 8.   Examples of cell stages as a cell passes from the pluripotent stem cell to a terminally differentiated state. In the case of heart, the progenitor cells are committed very early during differentiation, and there are no adult stem cells that can be used for cell replacement therapy of cardiac myocytes. One possible strategy to generate cardiac myocytes is to take pluripotent stem cells cultivated *in vitro*, differentiate them and select, by antibiotic selection or FACS sorting, a pure population of human cardiac cells with mitotic activity. If the cells can be appropriately synchronized and purified then cell replacement therapy in heart may be possible. While proof of principle has been performed in mouse, no long-term studies have been performed to ensure that transplantation of ES cell-derived donor cells do not give rise to tumors in heart. The aim is to find an alternative to organ transplantation to prolong life and improve life-style of an aged population increasingly at risk for cardiac-associated morbidity and mortality.

Before such a goal can be achieved, it will be important to determine what factors facilitate self-renewal and growth, promote differentiation to specific cells, and permit growth in the absence of feeder layers. At present, no one has successfully reported large-scale growth and cultivation, genetic manipulation or efficient cloning of human ES cells. Despite many years of research with mouse ES cells, much remains to be accomplished for other species, including human. Application of ES cell technology to pluripotent human ES cells also has enormous ethical implications, and international regulations are needed to prevent abuse and overcome legal problems. There are clearly enormous challenges for scientists and clinicians before any interventions can proceed, but the opportunities are vast. As long as this area of research continues to be supported by both government and private sources, potential benefits to man will be realized.

7.   References

1. Jazwinski, S.M. (1996) Science 273, 54-9.
2. Yancik, R. and Ries, L.A. (2000) Hematol. Oncol. Clin. North Am. 14, 17-23.
3. Yancik, R. (1997) Cancer 80, 1273-83.
4. Balducci, L. and Extermann, M. (2000) Hematol. Oncol. Clin. North Am. 14, 1-16.
5. Beaufrere, B. and Morio, B. (2000) Eur. J. Clin. Nutr. 54, S48-S53.
6. Fossel, M. (2000) *In Vivo* 14, 29-34.
7. Ladislas, R. (2000) Pathol Oncol Res 6, 3-9.
8. Wick, G., Jansen-Durr, P., Berger, P., Blasko, I. and Grubeck-Loebenstein, B. (2000) Vaccine 18, 1567-83.
9. Mattson, M.P., Pedersen, W.A., Duan, W., Culmsee, C. and Camandola, S. (1999) Ann. N.Y. Acad. Sci. 893, 154-75.
10. Schocken, D.D. (2000) Clin. Geriatr. Med. 16, 407-418.
11. Marin, J. and Rodriguez-Martinez, M.A. (1999) Exp. Gerontol. 34, 503-12.
12. Lakatta, E.G. (1999) in: W.R.Hazzard (Ed.), Principles of Geriatric Medicine and Gerontology, McGraw-Hill Companies, Inc., New York. pp. 645-660.
13. Task Force of the Working Group on Heart Failure of the European Society of Cardiology (1997) Eur Heart J. 18, 736-53.
14. Finch, C.E. and Tanzi, R.E. (1997) Science 278, 407-411.
15. Koh, G.Y., Soonpaa, M.H., Klug, M.G. and Field, L.J. (1995) J. Interv. Cardiol. 8, 387-93.
16. Soonpaa, M.H., Daud, A.I., Koh, G.Y., Klug, M.G., Kim, K.K., Wang, H. and Field, L.J. (1995) Ann. N.Y. Acad. Sci. 752, 446-54.
17. Platt, J.L. (1998) Nature 392, 11-7.
18. Gage, F.H. (1998) Nature 392, 18-24.
19. Boheler, K.R. and Fiszman, M. (1999) Cells Tissues Organs 165, 237-245.
20. Thomson, J.A., Itskovitz-Eldor, J., Shapiro, S.S., Waknitz, M.A., Swiergiel, J.J., Marshall, V.S. and Jones, J.M. (1998) Science 282, 1145-1147.
21. Shamblott, M.J. et al. (1998) Proc. Natl. Acad. Sci. USA 95, 13726-13731.
22. Reubinoff, B.E., Pera, M.F., Fong, C.Y., Trounson, A. and Bongso, A. (2000) Nat. Biotechnol. 18, 399-404.
23. Mountford, P., Nichols, J., Zevnik, B., O'Brien, C. and Smith, A. (1998) Reprod. Fertil. Dev. 10, 527-33.
24. Smith, A.G., Heath, J.K., Donaldson, D.D., Wong, G.G., Moreau, J., Stahl, M. and Rogers, D. (1988)

Nature 336, 688-690.

25. Doetschman, T.C., Eistetter, H., Katz, M., Schmidt, W. and Kemler, R. (1985) J.Embryol.Exp.Morph. 87, 27-45.

26. Brook, F.A. and Gardner, R.L. (1997) Proc. Natl. Acad. Sci. USA 94, 5709-5712.

27. Williams, R.L., Hilton,D.J., Pease,S., Willson,T.A., Stewart,C.L., Gearing,D.P., Wagner,E.F., Metcalf,D., Nicola,N.A., Gough,N.M. (1988) Nature 336, 684-687.

28. Nagy, A., Gocza, E., Diaz, E.M., Prideaux, V.R., Ivanyi, E., Markkula, M. and Rossant, J. (1990) Development 110, 815-21.

29. Nagy, A., Rossant, J., Nagy, R., Abramow-Newerly, W. and Roder, J.C. (1993) Proc. Natl. Acad. Sci. USA 90, 8424-8.

30. Evans, M.J. and Kaufman, M.H. (1981) Nature 292, 154-156.

31. Martin, G.R. (1981) Proc. Natl. Acad. Sci. USA 78, 7634-7638.

32. Wobus, A.M., Holzhausen, H., Jaekel, P. and Schoeneich, J. (1984) Exp.Cell.Res. 152, 212-219.

33. Guan, K., Rohwedel, J. and Wobus, A.M. (1999) Cytotechnology 30, 211-226.

34. Choi, K., Kennedy, M., Kazarov, A., Papadimitriou, J.C. and Keller, G. (1998) Development 125, 725-32.

35. Robertson, E.J. (1987) in: E.J. Robertson, (Ed.) Teratocarcinomas and embryonic stem cells - a practical approach, IRL Press Oxford, Washington, D.C. pp. 91-112.

36. Bradley, A., Evans, M., Kaufman, M.H. and Robertson, E. (1984) Nature 309, 255-6.

37. Thomas, K.R. and Capecchi, M.R. (1987) Cell 51, 503-12.

38. Rohwedel, J., Horak, V., Hebrok, M., Fuechtbauer, E.-M. and Wobus, A.M. (1995) Exp. Cell Res. 220, 92-100.

39. Prelle, K., Wobus, A.M., Krebs, O., Blum, W.F. and Wolf, E. (2000) Biochem. Biophys. Res. Commun. 277, 631-8.

40. Grepin, C., Nemer, G. and Nemer, M. (1997) Development 124, 2387-2395.

41. Braun, T. and Arnold, H.H. (1996) EMBO J. 15, 310-18.

42. Wobus, A.M. and Guan, K. (1998) Trends. Cardiovasc. Med. 8, 64-74.

43. Faessler, R. Rohwedel,J., Maltsev,V., Bloch,W., Lentini,S., Guan,K., Gullberg,D., Hescheler,J., Addicks,K., Wobus,A.M. (1996) J. Cell. Sci. 109, 2989-99.

44. Weitzer, G., Milner, D.J., Kim, J.U., Bradley, A. and Capetanaki, Y. (1995) Dev. Biol. 172, 422-39.

45. Smith, A.G., Nichols, J., Robertson, M. and Rathjen, P.D. (1992) Dev. Biol. 151, 339-51.

46. Lin, Q., Srivastava, D. and Olson, E.N. (1997) Cold Spring Harb. Symp. Quant. Biol. 62, 405-11.

47. Harvey, R.P. (1996) Dev. Biol. 178, 203-16.

48. Lin, Q., Schwarz, J., Bucana, C. and Olson, E.N. (1997) Science 276, 1404-7.

49. Wobus, A.M., Rohwedel, J., Maltsev, V. and Hescheler, J. (1997) Toxicology *in Vitro* 9, 477-488.

50. Barton, P.J.R., Boheler, K.R., Brand, N.J. and Thomas, P.S (1995) In: P.J.R. Barton, K.R. Boheler, N.J. Brand and P.S. Thomas (Eds.) Molecular Biology of Cardiac Development and Growth, R.G. Landes, Austin TX.

51. Olson, E.N. and Srivastava, D. (1996) Science 272, 671-676.

52. Ko, M.S. Kitchen, J.R., Wang, X., Threat, T.A., Hasegawa, A., Sun, T., Grahovac, M.J., Kargul, G.J., Lim, M.K., Cui, Y., Sano, Y., Tanaka, T., Liang, Y., Mason, S., Paonessa, P.D., Sauls, A.D., DePalma, G.E., Sharara, R., Rowe, L.B., Eppig, J., Morrell, C., Doi, H. (2000) Development 127, 1737-1749.

53. Anisimov, S.V., Tarasov, K.V., Riordon, D., Wobus, A.M. and Boheler, K.R. (submitted).

54. Parker, T.G. and Schneider, M.D. (1991) Ann. Rev. Physiol 53, 179-200.

55. Schneider, M.D., Roberts, R. and Parker, T.G. (1991) Mol.Biol.Med 8, 167-183.

56. Izumo, S., Nadal-Ginard, B. and Mahdavi, V. (1988) Proc. Natl. Acad. Sci. USA 85, 339-343.

57. Walsh, K. and Perlman, H. (1997) Curr. Opin. Genet. Dev. 7, 597-602.

58. Makino, S., Fukuda, K., Miyoshi, S., Konishi, R., Kodama, H., Pan, J., Sano, M., Takahashi, T., Hori, S. Abe, H., Hata, J-i., Umezawa, A., Ogawa, S. (1999) J. Clin. Invest. 103, 697-705.

60. Smith, A.G. and Hooper, M.L. (1987) Dev. Biol. 121, 1-9.

61. Burdon, T., Chambers, I., Stracey, C., Niwa, H. and Smith, A. (1999) Cells Tissues Organs 165, 131-43.

62. Matsuda, T., Nakamura, T., Nakao, K., Arai, T., Katsuki, M., Heike, T. and Yokota, T. (1999) EMBO J. 18, 4261-9.

63. Nichols, J., Zevnik, B., Anastassiadis, K., Niwa, H., Klewe-Nebenius, D. Chambers, I., Schoeler, H., Smith, A. (1998) Cell 95, 379-391.

64. Hirano, T., Nakajima, K. and Hibi, M. (1997) Cytokine Growth Factor Rev. 8, 241-52.

65. Lenardo, M.J., Staudt, L., Robbins, P., Kuang, A., Mulligan, R.C. and Baltimore, D. (1989) Science 243, 544-546.

66. Schoeler, H.R., Balling, R., Hatzopoulos, A.K., Suzuki, N. and Gruss, P. (1989) EMBO J. 8, 2551-7.

67. Schoeler, H.R., Ruppert, S., Suzuki, N., Chowdhury, K., Gruss, P. (1990) Nature 344, 435-439.

68. Burdon, T., Stracey, C., Chambers, I., Nichols, J. and Smith, A. (1999) Dev. Biol. 210, 30-43.

69. Pesce, M., Anastassiadis, K. and Schoeler, H.R. (1999) Cells Tissues Organs 165, 144-52.

70. Anisimov, S.V., Lakatta, E.G. and Boheler, K.R. (2001) Eur. J. Heart Failure 3, 271-281.

71. Edwards, M.K., Harris, J.F. and McBurney, M.W. (1983) Mol. Cell. Biol. 3, 2280-6.

72. Edwards, M.K. and McBurney, M.W. (1983) Dev. Biol. 98, 187-91.

73. Wobus, A.M., Kleppisch, T., Maltsev, V. and Hescheler, J. (1994) *In Vitro* Cell Dev. Biol. Anim. 30A, 425-34.

74. Velculescu, V.E., Zhang, L., Vogelstein, B. and Kinzler, K.W. (1995) Science 270, 484-7.

75. Velculescu, V.E. et al. (1999) Nat. Genet. 23, 387-8.

76. Dunwoodie, S.L., Rodriguez, T.A. and Beddington, R.S. (1998) Mech. Dev. 72, 27-40.

77. Jaradat, S.A., Tanaka, T.S., O'Neill, L., Chen, G., Banergee, N., Zhang, M.Z., Boheler, K.R. and Ko, M.S.H. (2001) Keystone Symposium abstracts (in press).

78. Srivastava, D. (2000) Circ. Res. 86, 917-8.

79. Robbins, J., Doetschman, T., Jones, W.K. and Sanchez, A. (1992) Trends Cardiovasc. Med. 2, 44-50.

80. Gossler, A., Joyner, A.L., Rossant, J. and Skarnes, W.C. (1989) Science 244, 463-5.

81. Friedrich, G. and Soriano, P. (1991) Genes Dev. 5, 1513-23.

82. Joyner, A.L., Auerbach, A. and Skarnes, W.C. (1992) Ciba. Found. Symp. 165, 277-88.

83. Skarnes, W.C., Auerbach, B.A. and Joyner, A.L. (1992) Genes Dev. 6, 903-18.

84. Wurst, W., Fukuda, K., Miyoshi, S., Konishi,R., Kodama, H., Pan, J., Sano, M., Takahashi, T., Hori, S., Abe, H., Hata, J-i., Umezawa, A., Ogawa, S. et al. (1995) Genetics 139, 889-99.

85. Forrester, L.M., Nagy, A., Sam, M., Watt, A., Stevenson, L., Bernstein, A., Joyner, A.L. and Wurst, W. (1996) Proc. Natl. Acad. Sci. USA 93, 1677-82.

86. Baker, R.K., Haendel, M.A., Swanson, B.J., Shambaugh, J.C., Micales, B.K. and Lyons, G.E. (1997) Dev. Biol. 185, 201-14.

87. Stanford, W.L., Caruana, G., Wallis, K.A., Inamdar, M., Hidaka, M., Bautch, V.L. and Bernstein, A. (1998) Blood 92, 4622-4631.

88. Motoyama, J., Kitajima, K., Kojima, M., Kondo, S. and Takeuchi, T. (1997) Mech. Dev. 66, 27-37.

89. Lee, Y., Song, A.J., Baker, R., Micales, B., Conway, S.J. and Lyons, G.E. (2000) Circ.Res. 86, 932-8.

90. Takeuchi, T., Kojima, M., Nakajima, K. and Kondo, S. (1999) Mech. Dev. 86, 29-38.

91. Toyoda, M., Kojima, M. and Takeuchi, T. (2000) Biochem. Biophys. Res. Commun. 274, 332-6.

92. Wobus, A.M., Guan, K., Yang, H.-T. and Boheler, K.R. (2002) in: K. Turksen (Ed.) Embryonic Stem

Cells, Methods in Molecular Biology, Humana Press, Totowa, New Jersey. Vol. 185, pp.127-156.

93. Wobus, A.M., Wallukat, G. and Hescheler, J. (1991) Differentiation 48, 173-182.

94. Wiles, M.V. and Keller, G. (1991) Development 111, 259-267.

95. Wobus, A.M., Rohwedel, J., Maltsev, V. and Hescheler, J. (1995) Ann. N.Y. Acad. Sci. 752, 460-9.

96. Hescheler, J., Fleischmann, B.K., Lentini, S., Maltsev, V.A., Rohwedel, J., Wobus, A.M. and Addicks, K. (1997) Cardiovasc. Res. 36, 149-62.

97. Maltsev, V.A., Rohwedel, J., Hescheler, J. and Wobus, A.M. (1993) Mech. Dev. 44, 41-50.

98. Maltsev, V.A., Wobus, A.M., Rohwedel, J., Bader, M. and Hescheler, J. (1994) Circ. Res. 75, 233-244.

99. Maltsev, V.A., Ji, G.J., Wobus, A.M., Fleischmann, B.K. and Hescheler, J. (1999) Circ Res 84, 136-145.

100. Sauer, H., Hofmann, C., Wartenberg, M., Wobus, A.M. and Hescheler, J. (1998) Exp. Cell. Res. 238, 13-22.

101. Wobus, A.M., Rohwedel, J., Maltsev, V. and Hescheler, J. (1994) Roux's Arch. Dev. Biol. 204, 36-45.

102. Hescheler, J., Fleischmann, B.K., Wartenberg, M., Bloch, W., Kolossov, E., Ji, G., Addicks, K. and Sauer, H. (1999) Cells Tissues Organs 165, 153-64.

103. Metzger, J.M., Samuelson, L.C., Rust, E.M. and Westfall, M.V. (1997) Trends Cardiovasc. Med. 7, 63-68.

104. Narita, N., Bielinska, M. and Wilson, D.B. (1996) Development 122, 3755-3764.

105. Guan, K., Fuerst, D.O. and Wobus, A.M. (1999) Eur. J. Cell. Biol. 78, 813-23.

106. van der Ven, P.F. and Fuerst, D.O. (1997) Cell Struct. Funct. 22, 163-71.

107. Auerbach, D., Rothen-Ruthishauser, B., Bantle, S., Leu, M., Ehler, E., Helfman, D. and Perriard, J.C. (1997) Cell Struct. Funct. 22, 139-46.

108. Ehler, E., Rothen, B.M., Hammerle, S.P., Komiyama, M. and Perriard, J.C. (1999) J. Cell. Sci. 112, 1529-39.

109. Metzger, J.M., Lin, W.-I. and Samuelson, L.C. (1996) Circ. Res. 78, 547-552.

110. Wobus, A.M., Kaomei, G., Shan, J., Wellner, M-C., Rohwedel, J., Guanju, J., Fleischmann, B., Katus, H.A., Hescheler, J., Franz, W-M. (1997) J. Mol. Cell. Cardiol. 29, 1525-39.

111. Kolossov, E., Fleischmann, B.K., Liu, Q., Bloch, W., Viatchenko-Karpinski, S., Manzke, O., Ji, G.J., Bohlen, H., Addicks, K., Hescheler, J. (1998) J. Cell. Biol.143, 2045-2056.

112. Meyer, N., Jaconi, M., Landopoulou, A., Fort, P. and Puceat, M. (2000) FEBS Lett. 478, 151-8.

113. Drab, M., Haller, H., Bychkov, R., Erdmann, B., Lindschau, C., Haase, H., Morano, I., Luft, F.C., Wobus, A.M. (1997) FASEB J. 11, 905-915.

114. Metzger, J.M., Lin, W.I., Samuelson, L.C. (1994) J. Cell. Biol. 126, 701-711.

115. Moorman, A.F.M. and Lamers, W.H. (1992) in: A. El Haj (Ed.) Molecular Biology of Muscle, The Company of Biologists Limited, Cambridge, pp. 285-300.

116. Moorman, A.F.M. and Lamers, W.H. (1994) Trends Cardiovasc. Med. 4, 257-264.

117. Moorman, A.F.M., Vermeulen, J.L.M., Koban, M.U., Schwartz, K., Lamers, W.H. and Boheler, K.R. (1995) Circ. Res. 76, 616-625.

118. Koban, M.U., Moorman, A.F., Holtz, J., Yacoub, M.H. and Boheler, K.R. (1998) Cardiovasc. Res. 37, 405-23.

119. Periasamy, M., Reed, T.D., Liu, L.H., Ji, Y., Loukianov, E., Paul, R.J., Nieman, M.L., Riddle, T., Duffy, J.J., Doetschman, T., Lorenz, J.N., Shull, G.E. (1999) J. Biol. Chem. 274, 2556-2562.

120. Takeshima, H., Komazaki, S., Hirose, K., Nishi, M., Noda, T. and Iino, M. (1998) EMBO J. 17, 3309-3316.

121. Xu, C., Liguori, G., Adamson, E.D. and Persico, M.G. (1998) Dev. Biol. 196, 237-47.

122. Yang, H.-T., Lakatta, E.G. and Boheler, K.R. (2000) Circulation 102 (suppl. II), II-141 (abstract).

123. Okabe, S., Forsberg-Nilsson, K., Spiro, A.C., Segal, M. and McKay, R.D. (1996) Mech. Dev. 59, 89-102.

124. Brustle, O. (1999) Brain Pathol. 9, 527-45.

125. Brustle, O., Jones, K.N., Learish, R.D., Karram, K., Choudhary, K., Wiestler, O.D., Duncan, I.D. and McKay, R.D. (1999) Science 285, 754-6.

126. Lee, S.H., Lumelsky, N., Studer, L., Auerbach, J.M. and McKay, R.D. (2000) Nat. Biotechnol. 18, 675-9.

127. Li, M., Pevny, L., Lovell-Badge, R. and Smith, A. (1998) Curr. Biol. 8, 971-974.

128. Dinsmore, J., Ratliff, J., Deacon, T., Pakzaban, P., Jacoby, D., Galpern, W. and Isacson, O. (1996) Cell Transplant. 5, 131-43.

129. Dinsmore, J., Ratliff, J., Jacoby, D., Wunderlich, M. and Lindberg, C. (1998) Theriogenology 49, 145-51.

130. Klug, M.G., Soonpaa, M.H., Koh, G.Y. and Field, L.J. (1996) J. Clin. Invest. 98, 216-224.

131. Muller, M., Fleischmann, B.K., Selbert, S., Ji, G.J., Endl, E., Middeler, G., Muller, O.J., Schlenke, P., Frese, S., Wobus, A.M., Hescheler, J., Katus, H.A., Franz, W.M. (2000) FASEB J. 14, 2540-8.

132. Capecchi, M.R. (1989) Science 244, 1288-92.

133. Capecchi, M.R. (1989) Trends Genet. 5, 70-6.

134. Rossant, J., Bernelot-Moens, C. and Nagy, A. (1993) Philos. Trans. R. Soc. Lond. B. Biol. Sci. 339, 207-15.

135. Sharp, M.G., Kantachuvesiri, S. and Mullins, J.J. (1997) Curr. Opin. Nephrol. Hypertens. 6, 51-7.

136. Mullins, L.J., Morley, S.D. and Mullins, J.J. (1996) J. Hum. Hypertens. 10, 627-31.

137. Chien, K.R. (1993) Science 260, 916-917.

138. Chien, K.R. (1995) Am. J. Physiol. 269, H755-66.

139. Christensen, G., Wang, Y. and Chien, K.R. (1997) Am. J.Physiol. 272, H2513-24.

140. Becker, K.D., Gottshall, K.R. and Chien, K.R. (1996) Hypertension 27, 495-501.

141. Chien, K.R. (1996) J. Clin. Invest. 97, 901-9.

142. Shamblott, M.J., Axelman, J., Littlefield, J.W., Blumenthal, P.D., Huggins, G.R., Cui, Y., Cheng, L. and Gearhart, J.D. (2001) Proc.Natl. Acad. Sci. U S A 98, 113-118.

143. Pera, M.F., Reubinoff, B. and Trounson, A. (2000) J. Cell Sci. 113, 5-10.

144. Bongso, T.A., Fong, C.Y., Ng, C.Y. and Ratnam, S.S. (1994) Cell. Biol. Int. 18, 1181-9.

145. Bongso, A., Fong, C.Y., Ng, S.C. and Ratnam, S. (1994) Hum.Reprod. 9, 2110-7.

146. Schuldiner, M., Yanuka, O., Itskovitz-Eldor, J., Melton, D.A. and Benvenisty, N. (2000) Proc. Natl. Acad. Sci. USA 97, 11307-12.

147. Itskovitz-Eldor, J., Schuldiner, M., Karsenti, D., Eden, A., Yanuka, O., Amit, M., Soreq, H. and Benvenisty, N. (2000) Mol. Med. 6, 88-95.

148. Koh, G.Y., Klug, M.G., Soonpaa, M.H. and Field, L.J. (1993) J. Clin. Invest. 92, 1548-1554.

149. Koh, G.Y., Soonpaa, M.H., Klug, M.G., Pride, H.P., Cooper, B.J., Zipes, D.P. and Field, L.J. (1995) J. Clin. Invest. 96, 2034-42.

150. Greiner, D.L., Hesselton, R.A. and Shultz, L.D. (1998) Stem Cells 16, 166-77.

151. Bianco, P., Kuznetsov, S.A., Riminucci, M., Fisher, L.W., Spiegel, A.M. and Robey, P.G. (1998) J. Clin. Invest. 101, 1737-44.

152. Fuchs, E. and Segre, J.A. (2000) Cell 100, 143-55.

153. Clarke, D.L., Johansson, C.B., Wilbertz, J., Veress, B., Nilsson, E., Karlstrom, H., Lendahl, U. and Frisen, J. (2000) Science 288, 1660-3.

154. Bjornson, C.R., Rietze, R.L., Reynolds, B.A., Magli, M.C. and Vescovi, A.L. (1999) Science 283, 534-7.

155. Gussoni, E., Soneoka, Y., Strickland, C.D., Buzney, E.A., Khan, M.K., Flint, A.F., Kunkel, L.M. and Mulligan, R.C. (1999) Nature 401, 390-4.

156. Petersen, B.E., Bowen, W. C., Patrene, K. D., Mars, W. M., Sullivan, A. K., Murase, N., Boggs, S. S., Greenberger, J. S., Goff, J. P. (1999) Science 284, 1168-70.

157. Galli, R., Borello, U., Gritti, A., Minasi, M. G., Bjornson, C., Coletta, M., Mora, M., De Angelis, M., G.,

Fiocco, R., Cossu, G., Vescovi, A. L. (2000) Nat. Neurosci. 3, 986-91.

158. Jackson, K.A., Mi, T. and Goodell, M.A. (1999) Proc. Natl. Acad. Sci. USA 96, 14482-6.

159. Fukuda, K. (2000) Tanpakushitsu Kakusan Koso. 45, 2078-84.

160. Watt, F.M. and Hogan, B.L. (2000) Science 287, 1427-30.

161. Campbell, K.H., McWhir, J., Ritchie, W.A. and Wilmut, I. (1996) Nature 380, 64-6.

162. Young, L.E., Sinclair, K.D. and Wilmut, I. (1998) Rev. Reprod. 3, 155-63.

163. Wilmut, I., Schnieke, A.E., McWhir, J., Kind, A.J. and Campbell, K.H. (1997) Nature 385, 810-813.

164. Rohwedel, J., Sehlmeyer, U., Shan, J., Meister, A. and Wobus, A.M. (1996) Cell Biology International 20, 579-587.

165. Wobus, A.M., Rohwedel, J., Struebing, C.,, Shan, J., Adler, K., Maltsev, V. and Hescheler, J. (1997) In: E. Klug and R. Thiel (Eds.), Methods in Developmental Toxicology and Biology, Blackwell Science, Berlin/Vienna, pp. 1-17.

CHAPTER 8

ADULT SKELETAL MUSCLE GROWTH AND REGENERATION: THE DEVELOPMENT AND DIFFERENTIATION OF MYOGENIC STEM CELLS

PATRICK SEALE and MICHAEL A. RUDNICKI

Table of contents

Stem Cells: A Cellular Fountain of Youth.   Ed. by Mark P. Mattson and Gary Van Zant. 177 — 200

1.   Introduction

Vertebrate skeletal muscle contains a population of mononuclear cells, termed satellite cells, which reside underneath the basal lamina of myofibers and function as stem cells in growing and regenerating postnatal muscles. The embryonic origin and ontogeny of satellite cells remains elusive, however recent work raises the possibility that they may arise from precursors associated with the developing vascular system. In addition to satellite cells, a distinct population of cells with multipotent stem cell activity resides within skeletal muscle tissue, so-called muscle-derived stem cells (MSC). Following intravenous injection into lethally irradiated mice, MSCs are recruited into regenerating skeletal myofibers and are able to reconstitute the hematopoietic system. In this review, we will focus on the biology of both muscle satellite cells and muscle-derived stem cells including the possibility of exploiting the properties of these cells in designing effective cell based therapies for degenerative diseases.

2.   Skeletal muscle satellite cells

Skeletal muscle satellite cells are defined on the basis of their morphology and anatomical location relative to mature myofibers. Satellite cells adhere to the surface of myotubes prior to the formation of the basal lamina (Figure 1A,B) [1-4]. They reside in grooves or depressions between the basal lamina and sarcolemma of mature fibers and make up 2-10% of sublaminar muscle nuclei depending on the type of fibers with which they associate [2]. The association of satellite cells with adjacent myofibers is mediated by cell-cell interactions involving the activity of M-Cadherin (muscle Cadherin), N-CAM (Neural Cell Adhesion Molecule) and other cell-adhesion molecules [5, 6]. Adult satellite cells are also characterized by high nuclear to cytoplasmic ratios, heterochromatic nuclei and a paucity of cytoplasmic organelles (Figure 1B) [4, 7].

At birth, satellite cells constitute the only myogenic cells in skeletal muscle with proliferative capacity [8]. Thus, the satellite cell compartment mediates postnatal muscle development and is the primary means by which the adult muscle mass is formed [8-10]. As such, the overall population of satellite cells decreases with increasing age in developing muscles [10-12]. At birth satellite cells account for about 32% of muscle nuclei followed by a drop to less than 5% in the adult (2 months for mice) [2]. The decline in satellite cell number as the postnatal muscle develops is a direct reflection of satellite cell fusion into new or pre-existing myofibers.

2.1.   Developmental origin of satellite cells

Satellite cells are distinct from the embryonic myogenic lineages and first appear in the limbs of mouse embryos at about 17.5 days post coitum (dpc) [2, 13-16]. The birth of satellite cells during vertebrate development constitutes the third wave of myogenesis following the formation of both embryonic and fetal myogenic cells [15, 16]. Satellite cells can be distinguished from these earlier myogenic lineages on the basis of: myosin

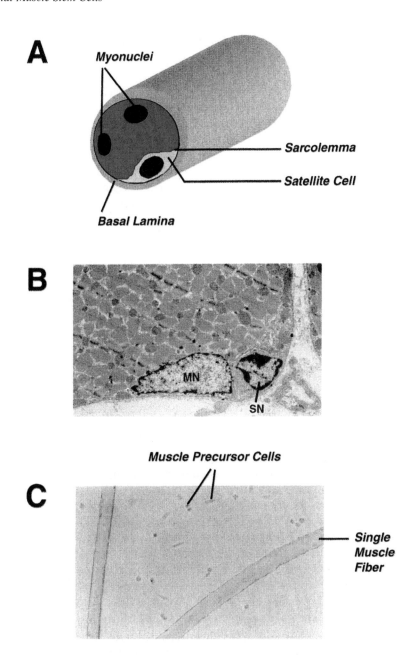

Figure 1. (A) Schematic representation of the anatomical position of satellite cells relative to mature myofibers. Satellite cells are mononuclear and are located in between the basal lamina and sarcolemma of mature myofibers. (B) Transmission electron micrograph of skeletal muscle depicting a satellite cell (SN) and fiber myonucleus (MN). Satellite cells possess densely staining heterochromatic nuclei and are characterized by organelle-poor cytoplasms. (C) Satellite cells within intact single muscle fibers give rise to daughter myogenic precursor cells following activation in culture.

heavy chain isoform expression; distinct morphological appearance; and a capacity to differentiate appropriately in the presence of the phorbol ester TPA in cell culture [2, 13-16].

Embryonic and fetal myoblasts are committed to the myogenic fate during somitogenesis in response to signals from adjacent tissues including the mesoderm and notochord [17-19], however the origin of satellite cells, which arise later in development, has been less carefully studied. Classical quail-chick chimerae experiments in which quail somitic mesoderm was grafted into 2-day chick embryos revealed the presence of quail satellite cells associated with host chick myofibers [4]. Based largely on this study, it has been assumed that satellite cells or their precursors originate from the myotome within the somite, as do earlier myogenic progenitors, however follow-up experiments designed to specifically address this issue have not been reported.

A recent study by DeAngelis et al. [13], has caught the attention of developmental biologists and challenges pre-conceived notions concerning the origin of satellite cells. Their report convincingly demonstrates that clonal satellite cell derived myogenic precursors are readily isolated from the embryonic dorsal aorta but not from the somite of mouse embryos at different developmental stages. Moreover, analogous myogenic precursors are found in the limbs of later-stage *c-Met-/-* and *Pax3-/-* mutant embryos that do not possess somite derived migratory myoblasts [20, 21]. Myoblasts isolated from these mutant mice may thus arise from blood vessels or associated cells within developing limb-buds. Based on this work, it is speculated that the differentiation of multipotential precursors associated with the embryonic vasculature occurs as a function of the tissue that is perfused, for example, vessels which colonize skeletal muscle, contain progenitors that give rise to satellite cells. Consistent with this hypothesis, aorta-derived myogenic cells express myogenic and endothelial markers that are also expressed in adult satellite cells [13].

Developing embryonic vessels are colonized by migratory populations of angioblasts which arise in the paraxial mesoderm (in somite) and differentiate from the mesoderm as solitary cells before fusing to form primitive blood vessels. [22-24]. The origin of blood vessels thus raises the possibility that a proportion of satellite cells are derived from progenitors in the paraxial mesoderm, which also give rise to endothelial cells. Furthermore, aortic precursor cells, which form the primitive dorsal aortae, are closely associated with the ventral surface of the somite [23]. The anatomical proximity of vasculogenic progenitors to the somite suggests that the explants used in experiments by Armand [4], may have contained aortic precursors with the capacity to differentiate as satellite cells. Further studies are required to characterize the precise source of myogenic progenitors derived from embryonic vascular structures. The experiments by de Angelis et al. suggest the possibility that myogenic cells can be derived from pericytes (specialized vessel supporting cells), blood vessel endothelial cells or circulating endothelial cells [25]. Collectively these observations support a model for satellite cell development, which occurs independently of myogenic specification within the somite [13, 26].

2.2. Satellite cells in adult muscle

Satellite cells in adult skeletal muscle are normally mitotically quiescent but are activated (i.e. initiate multiple rounds of proliferation) in response to a variety of stimuli including weight bearing exercise, trauma, stretching and denervation [12, 27-30]. In response to signals in regenerating muscle, satellite cells can cross the myofiber basal lamina and migrate to distal sites of injury [31, 32]. The descendants of activated satellite cells, called myogenic precursor cells (MPCs), undergo multiple rounds of division prior to fusing with existing or new myofibers (Figure 1C) [12, 33]. It is important to stress the non-equivalence of satellite cells and their daughter MPCs as determined on the basis of various biological and biochemical criteria. By definition a satellite cell is quiescent, does not express muscle determination genes and resides beneath the basal lamina of intact fibers. By contrast, MPCs are determined myogenic cells, which express a wide array of muscle specific transcription factors and structural genes.

The essential role of satellite cells in muscle regeneration, muscle hypertrophy and post-natal muscle growth is well documented. However, an understanding of the molecular pathways that regulate the activation and function of myogenic stem cells remains poorly understood. Nevertheless, key insights regarding the function of the myogenic regulatory factors (MRFs) and the paired-box containing transcription factor, Pax7 have provided insight into the mechanisms regulating the development, activation and differentiation of muscle satellite cells.

2.2.1. *The Myogenic regulatory factors (MRFs)*

The MyoD-family of bHLH transcription factors (MRFs) is required for the commitment and differentiation of embryonic myoblasts during development. The primary MRFs, Myf5 and MyoD are required for the determination of myoblasts from progenitors in the somite in response to signals from the neural tube and paraxial mesoderm respectively [18]. Although, MyoD and Myf5 possess overlapping roles in embryonic development, hypaxial muscle development requires the initial activation of *MyoD* whereas epaxial muscle formation is dependent on *Myf5* [34-36]. The secondary MRFs, Myogenin and MRF4 function downstream to regulate terminal differentiation [37, 38].

The MRF expression program during satellite-cell activation, proliferation, and differentiation appears analogous to the program manifested during the embryonic development of skeletal muscle. Quiescent satellite cells are believed to express no detectable levels of MRFs [33, 39, 40]. MyoD or Myf5 is rapidly up regulated within 3 hours of experimentally induced muscle injury prior to expression of Proliferating Cell Nuclear Antigen (PCNA), a marker for cell proliferation (Figure 2) [39]. However, the presence of LacZ positive satellite cells in freshly isolated muscle fibers from *Myf5nLacZ* mice (*LacZ* expressed from *Myf5* locus) has been suggested to indicate expression of *Myf5* in quiescent cells [41, 42]. It remains to be determined whether satellite cell expression of *Myf5nLacZ* is allele-specific or whether single muscle fiber preparation results in low level expression. Myogenin is expressed last during the time associated with myoblast fusion and differentiation (Figure 2) [33, 40].

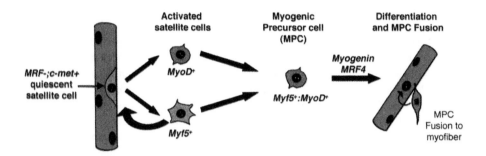

Figure 2.   Myogenic Regulatory Factor Activity in muscle satellite cell activation and differentiation. Quiescent satellite cells expressing *c-Met* but no MRFs give rise to activated satellite cells initially expressing either *MyoD* or *Myf5* followed by a developmental stage in which *MyoD* and *Myf5* are co-expressed in myogenic precursor cells. Activated satellite cells expressing *Myf5* but not *MyoD* appear to have increased propensity for self-renewal and may function to replenish the pool of quiescent stem cells *in vivo*. Terminal differentiation of MPCs requires the activity of the secondary MRFs, Myogenin and MRF4.

Analysis of gene expression by RT-PCR within single satellite cells on intact myofibers showed expression of *c-Met* (receptor for Hepatocyte Growth Factor) but no detectable levels of MRF mRNA in quiescent cells [43]. Interestingly, activated satellite cells (satellite cells entering the cell cycle) first express either *Myf5* or *MyoD* followed by a developmental stage in which *Myf5* and *MyoD* are co-expressed (Figure 2). Following proliferation, *myogenin* and *MRF4* are expressed in cells beginning their differentiation program. *In vivo* analysis of MRF expression by immunohistochemistry in regenerating muscle shows that 50% of activated satellite cells co-express MyoD and Myf-5, 30% express MyoD alone and 20% express Myf5 alone 3 hours post injury [39]. Collectively these results suggest independent entries into the satellite cell developmental program, either by activation of *MyoD* or *Myf5*, analogous to the kinetics of MRF expression in embryonic myoblasts that give rise to the MyoD-dependent hypaxial and Myf5-dependent epaxial muscles [44].

2.2.2.   *Requirement for MyoD in satellite cell function*

To investigate the role of MyoD in satellite cell function, *MyoD-/-* mice [45] were interbred with *mdx* mice. The *mdx* mice carry a loss-of-function point mutation in the X-linked *dystrophin* gene, and thus represent an animal model for human Duchenne and Becker muscular dystrophy [46, 47]. The compound mutant mice exhibit markedly increased penetrance of the *mdx* phenotype characterized by muscle atrophy and increased myopathy leading to premature death [48]. Skeletal muscle from *MyoD-/-* mice displays a strikingly reduced capacity for regeneration following experimentally induced injury despite increased numbers of satellite cells [48]. Consistent with these observations, single muscle fibers isolated from *MyoD-/-* mice display abnormal branching consistent with chronic muscle regeneration *in vivo* [49]. In the absence of

MyoD, myogenic progenitors undergo an apparent increase in numbers as a consequence of an increased propensity for self-renewal rather than progression through their differentiation program [48].

Further analysis of satellite cell-derived primary cultures from *MyoD-/-* hind limb muscle revealed a profound differentiation deficit [49-51]. Under conditions that normally induce differentiation of wildtype myoblasts, *MyoD-/-* cells continue to proliferate and only after several days yield reduced numbers of predominantly mononuclear myotubes. The observed differentiation deficit may result from an inability to activate appropriate levels of MRF4 mRNA [49]. Taken together, data from *in vivo* and *in vitro* experiments suggest that *MyoD-/-* myogenic cells represent an intermediate stage between a quiescent satellite cell and a MPC (Figure 2) [50]. A definitive role for Myf5 in satellite cell activation has yet to be established since it has not been possible to analyze muscle regeneration in *Myf5-/-* mice, which die perinatally due to inadequate rib development [52]. However, analysis of postnatal muscle from viable *Myf5-/-* mice recently generated by Kaul [53], will elucidate the role of Myf5 in satellite cell activation and differentiation.

2.2.3.   *Pax7 is required for Myogenic satellite cell development*

Pax7 belongs to a family of genes that encodes paired-box containing transcription factors involved in the control of diverse developmental processes [54-56]. In addition to its role in the development of neural crest cells, Pax7 is required for the specification of the muscle satellite cell lineage [57, 58]. The skeletal muscles of *Pax7* deficient mice are histologically indistinguishable from wild type muscle at birth however they are completely devoid of muscle satellite cells. The normal formation of embryonic muscle in *Pax7* deficient animals underscores the assertion that embryonic myoblasts and satellite cells develop independently of one another.

The specific requirement for Pax7 in satellite cell ontogeny is paralleled by the indispensable function of its paralogous gene family member, Pax3 in the developmental program of embryonic myoblasts. Pax3 functions genetically upstream of MyoD to promote expansion of muscle precursors in the somite and is required for the migration of myoblasts to developing limb buds [59-62]. Furthermore, ectopic expression of Pax3 in mesodermal explants is sufficient to activate skeletal myogenesis [63].

The molecular mechanism by which Pax3 specifies embryonic myogenic progenitors has not been resolved. Pax3 has however been suggested to function together with the homeodomain containing protein Six1, and the putative transcription factors Dach2 and Eya2 to regulate expansion of pre-muscle cell masses in the somite [64]. Interestingly, a homologous genetic network functions together with *eyeless* (*Drosophila* homolog of Pax6) to specify compound eye development in *Drosophila* [65]. Collectively these results suggest the possibility that different Pax factors specify diverse cell lineages using a well-conserved molecular pathway.

Based on the high degree of similarity between the Pax3 and Pax7 proteins, it is tempting to speculate that similar genetic mechanisms function to regulate satellite cell development. However, the expression patterns of Dach2, Eya2, and Six1 have yet to be analyzed in the context of adult myogenesis. An important area of future investigation will be to identify both upstream factors, which regulate Pax expression in presumptive muscle precursors as well as downstream target genes, which mediate myogenic specialization. Analysis of Pax7 gene expression in the latter stages of mouse or chick development may also provide insight into the developmental origin of satellite cells.

2.3.   Satellite cells as stem cells

Satellite cells are often described as stem cells due to their capacity for self-renewal, their large potential for expansion and their ability to generate daughter cells that differentiate as determined myogenic precursor cells [66, 67]. The absence of MRF mRNA in satellite cells prior to activation is consistent with the hypothesis that satellite cells represent a stem cell with an identity distinct from myoblasts. It is interesting therefore to speculate that the *de novo* activation of *Myf5* and *MyoD* transcription occurs in response to inductive signals analogous to those that occur during the specification of the myogenic lineage during embryonic development [19, 68].

Although activated satellite cells are restricted to the development of determined muscle cells *in vivo*, satellite cells within intact muscle fibers have the ability to activate alternate developmental programs in cell culture (Atsushi Asakura and Michael Rudnicki: unpublished observations). Furthermore, satellite cell derived myoblasts transdifferentiate to adipocytes or osteocytes in response to different culture conditions [69-71]. These results highlight a degree of plasticity for the satellite cell myogenic lineage, and demonstrate the importance of the *in vivo* microenvironment in promoting efficient myogenic differentiation of satellite cells during skeletal muscle growth and regeneration.

2.3.1.   *Satellite cell self renewal*

The number of quiescent satellite cells in adult muscle remains relatively constant over multiple cycles of degeneration and regeneration, demonstrating their inherent capacity for self-renewal [11, 72]. In addition, the relative number of satellite cells remains constant between adult (2 months) and old mice (> 2 years) [73]. By investigating the replication of satellite cells in immature postnatal muscles, Schultz [10] identified a population of "reserve" satellite cells that divide more slowly and remain largely in the G_0 phase of the cell cycle. It is hypothesized that reserve satellite cells function primarily to maintain a pool of quiescent stem cells, which are maintained into adulthood and old age, whereas the more rapidly dividing, metabolically active cells fuse to growing myofibers.

The concept of reserve muscle cells has also been described in cell culture experiments. Non-fusing myogenic cells (i.e. mononuclear) within differentiating muscle cultures can be cloned and initiate the growth of fusion competent myoblasts following re-stimulation with serum. Subsequent differentiation of these cultures again give rise to a discrete population of non-fusing reserve cells [74-76]. Interestingly, *MyoD* and *Myf5* are down regulated following mitogen withdrawal in non-fusing reserve cells suggesting that they have acquired a stem cell phenotype more closely resembling *in vivo* quiescent satellite cells [75].

The mechanisms responsible for satellite cell self-renewal in adult muscle are unknown however they may involve asymmetric cell division of a multipotent satellite cell, which divides giving rise to a committed MPC and a repopulating "self". Alternatively, activated satellite cells that express *Myf5* but not *MyoD* (i.e. Myf5⁺MyoD⁻ cell) may exist transiently in regenerating muscle and function primarily for self-renewal (Figure 2). This mechanism is postulated based on the phenotype of *MyoD-/-* satellite cells, which express high levels of *Myf5* but have reduced differentiation capacity and increased propensity for self-renewal. A third possibility would involve the de-differentiation of committed MPC to satellite cells. Interestingly, over expression of the homeobox containing msx1 transcription factor in terminally differentiated C2C12 myotubes can cause a small number of nuclei within myotubes to de-differentiate, and re-enter the cell cycle thereby generating viable mononuclear myoblasts [77]. Lastly, the possibility that muscle-derived stem cells function to replenish the satellite cell compartment is discussed in a later section (see also Figure 5). A more complete understanding of the mechanisms responsible for the self-renewal of satellite cells *in vivo* may lead to the development of novel therapeutic approaches to promote muscle cell replacement.

2.4.   Regulation of satellite cell activation and expansion

Satellite cells are activated in response to both modest and severe stressors suggesting perhaps multiple mechanisms for satellite cell activation. The physiological triggers that lead to satellite cell activation have yet to be defined. However a number of downstream molecules and mechanisms have been studied in regard to both activation of quiescent satellite cells and MPC expansion (Figure 3). In this section we will focus mainly on the ascribed functions of Hepatocyte Growth Factor (HGF), the Insulin-like Growth Factors (IGFs); members of the Fibroblast Growth Factor (FGF) family as well and the role of inflammation during muscle repair.

2.4.1.   Hepatocyte Growth Factor (HGF)

Although the physiological stimuli that trigger satellite cell activation are largely unknown, experimental evidence suggests the involvement of HGF in this process (Figure 3). HGF is a potent mitogen and chemotactic agent for satellite cells both *in vivo* and *in vitro* [78-81]. Furthermore, experimentally induced muscle damage results in the rapid co-localization of HGF and c-Met *in vivo* [82]. Finally, intramuscular infusion of HGF, followed by BrdU injection, results in increased recovery of BrdU positive MPCs after 30 hours in culture. Taken together these results suggest that HGF directly induces the activation of satellite cells [82]. The chemotactic activity of HGF suggests that it not only activates satellite cells but also induces their migration to sites of muscle damage [30-32, 80, 83].

Interestingly, administration of exogenous HGF into regenerating muscle is only able to increase the size of the MPC pool for up to one day following injury [84]. Surprisingly, exogenous HGF actually reduces the overall efficiency of muscle regeneration [84, 85]. These results imply that HGF may serve to activate satellite

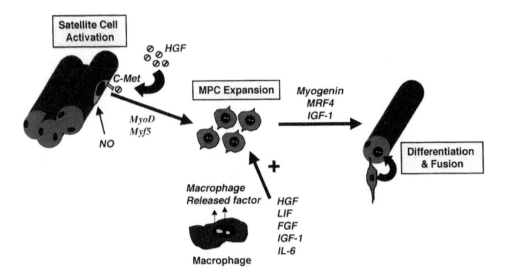

Figure 3.   Satellite cells are activated in response to physiological stimuli. The mechanism for satellite cell activation likely requires the activity of HGF, which binds its cognate receptor, c-Met, expressed on quiescent satellite cells. Release of Nitric Oxide (NO) from damaged fibers has also been suggested to activate satellite cells. Following activation, satellite cells give rise to Myogenic Precursor Cells (MPC), which undergo multiple rounds of division in response to a variety of growth factors. Following expansion of MPCs, they differentiate and fuse into regenerating or new myofibers.

cells but not to affect MPC proliferation during muscle regeneration. HGF may also directly inhibit terminal differentiation of MPCs thereby preventing fusion to new or pre-existing myofibers [86].

The source of HGF in regenerating muscle remains to be identified. A report by Gal-Levi et al. [85], demonstrates reciprocal expression of *c-Met* and *HGF* in growing myoblasts and myotubes i.e. myoblasts express *c-Met* alone, whereas newly formed myotubes express *HGF* alone. This observation implies that intact myotubes secrete HGF, which then acts in a paracrine fashion to promote satellite cell activation and proliferation. However, others have demonstrated expression of both *c-Met* and *HGF* in proliferating myoblasts in establishing a functional autocrine loop for myoblast expansion [79, 87]. HGF may also be released from the extracellular matrix through damage to the basal lamina. Although a mechanism for secretion of HGF in damaged muscle is not established, there is strong evidence to suggest that HGF/c-Met signal transduction has an important role for the repair of skeletal muscle.

2.4.2.   *Fibroblast Growth Factors (FGFs)*

Several members of the FGF family are expressed in developing skeletal muscle including FGF-1, 2, 4 5, 6, 7, 8 [88]. In addition, FGFR-1 and -4 (FGF-Receptor-1,-4) are expressed in quiescent muscle satellite cells suggesting a role for FGF signaling in directly affecting satellite cell activation or regulating quiescence [43]. FGF-2 (bFGF) positively regulates myoblast proliferation and also inhibits terminal differentiation and myotube formation in culture (Figure 3) [89-93]. Furthermore, expression levels of *FGF-2* in damaged mouse muscle correlates with the regenerative efficiency observed in different genetic backgrounds [94]. Interestingly, infusion of FGF-2 into regenerating muscle stimulates a significant increase in muscle IGF-1 concentration and total DNA but does not lead to increased protein concentration [95]. These results suggest that excess FGF-2 antagonizes the differentiation of myogenic cells thereby preventing their fusion into myofibers.

A role for Fibroblast Growth Factor-6 (FGF-6) in regulating either the proliferation or differentiation of MPCs is proposed based on the muscle regeneration deficit in mice carrying a targeted null mutation in *FGF-6* [96]. This work suggests that FGF-6 positively regulates MPC expansion during muscle regeneration. A recent report by Fiore et al., [97] however demonstrates no impairment in satellite cell function in independently derived *FGF6-/-* mice. It is possible that different genetic backgrounds and compensation by other FGFs may explain the differing results. The extent to which other members of the FGF family participate in postnatal muscle growth and regeneration remains to be analyzed.

2.4.3. *Insulin-like Growth Factors (IGFs)*

Insulin-like Growth Factors (IGF-1 and IGF-2) are implicated as mitogens, survival factors as well as differentiation factors for a variety of diverse cell lineages [98]. Evidence for the role of IGFs in skeletal muscle is exemplified in transgenic mice lacking the IGF-1-Receptor (IGF-1R) or deficient in IGF-1 and IGF-2, which display severe muscle hypotrophy resulting in premature death [99-101]. Furthermore, IGF-1 concentrations increase in muscle undergoing a hypertrophic response in response to experimentally induced functional overload [29, 102-104]. Consistent with a specific requirement for IGF-1 in myofiber growth, direct infusion of IGF-1 into the tibialis anterior muscles of adult rats is sufficient to induce a hypertrophic response (increase in protein content and size of myofibers). The IGF-1 induced hypertrophy leads to increased total muscle protein as well as increased DNA content indicating that the response is at least partially mediated by recruitment of satellite cell nuclei to the muscle [29, 102]. Conversely, intravenous administration of IGF-1 had no effect on skeletal muscle demonstrating the requirement for local IGF-1 production during regenerative and hypertrophic responses.

Recently, IGF-1 has been demonstrated to induce the Calcineurin-NFAT (Nuclear Factor of Activated T cells) signaling pathway leading to activation of GATA-2, a transcription factor whose up regulation is associated with myofiber hypertrophy [105-107]. Additionally, IGF-1 has been shown to promote myoblast survival by two independent PI3' Kinase (Phosphatidyl-Inositol-3' Kinase) pathways that both result in increased expression of the *p21* cdk inhibitor [108, 109]. Therefore, IGF-1 likely stimulates both MPC proliferation, and increased myofiber protein synthesis during muscle hypertrophy (Figure 3). Whether IGF-1 directly activates satellite cells or alternatively whether IGF-1 induced anabolic effects stimulate satellite cells via HGF or FGFs remains to be determined.

2.4.4. *Nitric Oxide (NO)*

A role for Nitric Oxide (NO) mediated satellite cell activation has recently been described [110]. By this model, bioactive NO, released locally from the membrane bound Dystrophin complex within damaged myofibers, induces the rapid activation of nearby satellite cells (Figure 3). Consistent with this model is the observation that inhibition of NOS (Nitric Oxide Synthase) activity was detrimental to experimentally induced regeneration [110]. However the normal histological appearance of NOS deficient muscle suggests that this mechanism is not absolutely required. Further examination of the role of NO signaling in satellite cell function is thus warranted.

2.4.5. *The inflammatory response*

Polymorphonuclear lymphocytes (PMNL) and macrophages (activated monocytes) migrate to sites of tissue damage within a few hours after trauma to muscle. Macrophages however, are the dominant immune cells present within regenerating muscle at 48 hours post injury [111, 112]. The role of macrophages in muscle regeneration is two fold in that macrophages phagocytose necrotic cell debris as well as secrete a soluble growth factor (yet to be characterized), which exerts a specific mitogenic effect on myoblasts (Figure 3) [113-115]. A vital role for macrophages in muscle regeneration is supported by the observation that myogenesis is markedly impaired in the absence of monocyte/macrophage infiltration [116].

The cytokines IL-6 (Interleukin-6) and LIF (Leukemia Inhibitory Factor) stimulate the proliferation of MPCs in culture (Figure 3) [117, 118]. In contrast to the late induction of *IL-6*, *LIF* expression is markedly increased 3 hours after muscle injury [118]. Moreover *LIF* mRNA is expressed in cultured myoblasts [118], suggesting that damaged muscle secretes LIF prior to infiltration of immune cells. Importantly, *LIF*-deficient mice display a markedly reduced capacity for muscle regeneration demonstrating that secretion of LIF in damaged muscles improves myofiber repair [119]. Continuous infusion of LIF to the dystrophic diaphragm of *mdx* mice results in reduced degeneration and increased myofiber diameter demonstrating a potential clinical application for LIF in promoting muscle growth and repair [120].

A role for leukocytes in satellite cell activation has also been proposed based on the observation that quiescent satellite cells express Vascular Cell Adhesion Molecule-1 (VCAM1)(a cell surface integrin molecule), whereas infiltrating leukocytes express the specific co-receptor VLA-4 (Integrin α4β1) [121-123]. Based on these observations a model is suggested in which cell-cell interactions mediated by VCAM1/VLA-4 ligation initiates genetic responses within satellite cells and immune cells to promote regeneration [121].

3.   Adult stem cells

3.1.   Myogenic precursors in bone-marrow

Recent studies have demonstrated the plasticity of stem cells derived from a variety of adult tissues [66]. A study by Ferrari et al., [124] first demonstrated the ability of bone marrow cells to give rise to myogenic cells that participate in muscle regeneration. In this study, bone marrow cells were isolated from mice carrying a muscle specific LacZ transgene and subsequently transplanted into immunodeficient recipients. Similarly, Bittner *et al.* [125], demonstrated recruitment of donor derived bone-marrow cells to skeletal and cardiac muscles of *mdx* mice. Donor male nuclei from marrow (detected by Y-chromosome specific fluorescence *in situ* hybridization (FISH)) are detected in endothelial cells in addition to muscle cells. Collectively these studies indicate the capacity of bone marrow derived stem cells to give rise to cells of myogenic and other lineages. Furthermore, the ability of these cells to be transported to sites of regeneration via the circulation suggests an effective strategy for the disseminated delivery of myogenic cells in the treatment of muscular dystrophies. However, in these studies there was no indication that the grafted cells made any contribution to the satellite cell compartment suggesting that donor bone marrow cells can only undergo terminal myogenic differentiation.

A subsequent study demonstrated that hematopoietic stem cells (HSCs) purified by FACS (fluorescence activated cell sorting) contribute nuclei to regenerating muscle fibers following intravenous injection into irradiated *mdx* mice [126]. The FACS method of isolating HSCs (also called side population or SP cells) depends on the principal that marrow-derived stem cells stain poorly with Hoechst 33342 due to high activity of mdr (multi-drug resistant)-like proteins [127, 128]. This procedure has been used to successfully purify HSC from many organisms including humans [128]. The SP cell population is heterogeneous and the identity or proportion of cells responsible for the observed stem cell activity remains unclear.

Importantly, following intravenous injection of 2000-5000 marrow SP cells into irradiated *mdx* mice, Dystrophin expression was restored in about 4% of the muscle fibers 12 weeks after engraftment. Similar to the results of Ferrari *et* al., [124] and Bittner *et al.*, [125] no donor-derived male nuclei were observed to segregate into the satellite cell compartment.

3.2.   Muscle-Derived Stem Cells (MSCs)

Use of the Hoechst/FACS stem cell isolation procedure revealed that skeletal muscle similarly possesses a robust SP fraction (Figure 4) [58, 126, 129]. Following intravenous injection of fractionated muscle-SP cells or whole muscle cells into irradiated mice, donor muscle derived cells were engrafted into the host bone marrow [126, 129]. Moreover, dissociated cells from skeletal muscle contribute to all major blood lineages three months following engraftment (Figure 4) [129]. Interestingly, MSCs (enriched in the muscle SP fraction) are recruited to regenerating myofibers more efficiently than HSCs with donor-derived nuclei detected in up to 9% of recipient muscle fibers [126]. Perhaps more significantly, MSCs apparently gave rise to satellite cells since Y-chromosome bearing donor cells were found between the Dystrophin positive sarcolemma and Laminin containing basal lamina in recently regenerated host fibers following transplantation [126]. Further experiments demonstrating co-localization of satellite cell specific markers and donor nuclei are required to directly address this question. Taken together, these experiments unequivocally demonstrate that muscle tissue contains a population of multipotent stem cells.

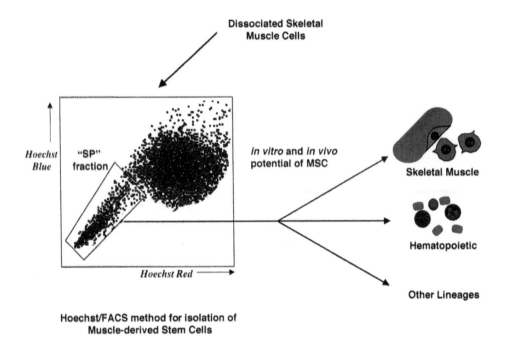

Figure 4.   Skeletal muscle contains a large SP fraction, which is isolated by FACS on the basis of Hoechst 33342-dye exclusion. The SP fraction possesses cells with stem cell activity, Muscle-Derived Stem Cells (MSC), that give rise to muscle cells hematopoietic cells and other cell types both *in vivo* and *in vitro*.

The study of adult stem cells including their remarkable plasticity has been an intense area of investigation since these reports were published. For example: neural stem cells (NSCs) can repopulate the bone marrow/blood system in lethally irradiated mouse recipients [130]; transplanted adult bone marrow gives rise to differentiated neuronal cell types [131]; marrow-derived cells differentiate into macroglia and microglia following injection into brain [132]; purified HSCs differentiate as liver cells following intravenous injection [133, 134] and bone marrow cells can contribute to bone, cartilage and lung formation in lethally irradiated recipients [135]. Therefore, these studies challenge the widely held view that tissue-specific stem cells are pre-determined to give rise to a particular cell type or cell types. In fact these stem cells may be pluripotent or even totipotent, possessing the ability to activate various genetic programs when exposed to the appropriate environmental cues. Thus pluripotential stem cells give rise to different cell types as a function of the growth factors and signals provided by their host tissue [136].

The low efficiency for the myogenic conversion of MSCs in cell culture implies that regenerating muscle provides the requisite micro-environmental cues to permit efficient myogenic specification. Recently, small numbers of cultured neural stem cells (neurospheres) were shown to differentiate as skeletal myogenic cells following exposure to myoblasts in coculture experiments [137]. This study suggests that cell-cell interactions and growth factors secreted by myoblasts are sufficient to allow neural stem cells to adopt a myogenic fate. Identification of the cell-surface molecules and soluble protein factors that drive stem cells along different developmental pathways may provide entirely new approaches to stimulate regenerative processes and treat degenerative diseases.

3.2.1. *Can muscle-derived stem cells give rise to satellite cells*

The identity of the MSC that resides in skeletal muscle tissue has yet to be defined. Initially, satellite cells were considered as the most likely source, however experimental evidence suggests otherwise. The observed presence of stem cells capable of robust hematopoietic differentiation in *Pax7* mutant muscle cultures, which lack satellite cells, clearly demonstrates that MSCs and satellite cells are distinct entities [58]. Furthermore, muscle satellite cells associated with single muscle fibers are incapable of hematopoietic colony formation under the same culture conditions that promotes hematopoietic differentiation of dissociated whole-muscle cells (Atsushi Asakura and Michael Rudnicki: unpublished data). Skeletal muscle is a complex tissue, which contains many cell types associated with connective tissues, nerves, and vasculature. It may be that the vascular associated progenitors that give rise to myogenic clones in embryonic dorsal aortic explants [13] persist in adult tissues thus giving rise to tissue-specific stem cells.

Several lines of evidence suggest that MSC may have the capacity to form satellite cells. Preliminary experiments by Gussoni et al. [126], identified donor derived cells in satellite cell positions within the host myofibers. Furthermore, MSCs can convert to myoblasts in cell culture experiments [126] (and our unpublished observations). Lastly, the observed increase in hematopoietic colony formation in cultures of *Pax7* mutant muscle cells suggests that MSCs normally give rise to satellite cells by a Pax7 dependent mechanism [58]. *Pax7* mRNA is not expressed by RT-PCR analysis in fractionated SP cells (unpublished observation). We thus predict that induction of Pax7 activity in MSCs induces myogenic specification by restricting alternative developmental pathways (Figure 5). This activity would be analogous to the established role for Pax5 in B-cell commitment [138, 139]. The model described here also suggests that satellite cells may be formed in a continuous fashion throughout life by recruitment of progenitor MSCs from within the SP fraction. The possibility that MSCs can directly give rise to myogenic cells in a Pax3/7 independent manner through direct activation of Myf5 similar to Shh (Sonic Hedgehog) induced epaxial muscle specification [18] cannot be discounted (Figure 5). In this way, MSCs may have the potential to fuse directly into regenerating fibers thereby bypassing the satellite cell compartment.

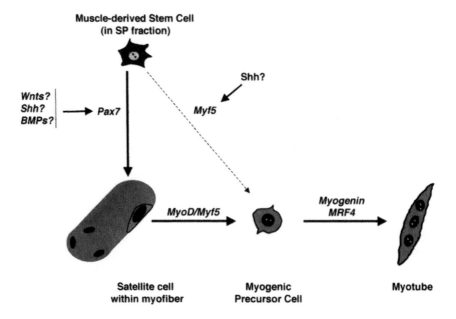

Figure 5.   Muscle-Derived Stem Cells (MSC) may give rise to myogenic precursor cells in a Pax7 dependent or independent manner. Activation of Pax7 in MSCs in response to local inductive signals may promote satellite cell formation both during embryonic development and in adult muscle to maintain a pool of quiescent cells. Alternatively or additionally, MSCs may directly differentiate as muscle cells in response to inductive signals such as Shh (Sonic Hedgehog) leading to the expression of Myf5 and direct entry into the myogenic program.

4.   Aging and Muscle Loss

The natural process of aging results in a progressive loss of muscle mass and strength resulting from the degeneration of myofibers possibly in response to neurodegeneration [140, 141]. Between the ages of 30 and 80 in humans it is estimated that there is a reduction of ~30% in muscle mass with a selective loss of fast fiber types [142]. It is also clear that old muscle is also more susceptible to injury and incomplete regeneration following exercise [143, 144]. Related to loss of muscle mass and function is the reduced activity of the growth hormone (GH)/IGF-1 axis and an increase in muscle catabolism [140].

The replicative capacity of MPCs from old muscle is seemingly intact in spite of an increased lag phase prior to the onset of DNA synthesis relative to cells from younger hosts [73, 145-147]. Furthermore, the stability of telomeres in satellite cells from young and old human muscle alike indicates that satellite cells undergo minimal turnover throughout adulthood and further suggests that a competent satellite cell population is present in aging muscles [148]. Moreover, the relative proportion of satellite cells present in old muscle however remains stable [73]. The inefficient regenerative processes in old muscle may thus result from alterations to *in vivo* microenvironmental components required for efficient activation and subsequent differentiation of satellite cells [149, 150]. Thus the degeneration of aging muscle seemingly reflects an inability to activate resident satellite cells rather than a defective stem cell compartment.

Strategies to prevent the aging associated loss in muscle mass is an active area of investigation. Several groups report protection against muscle loss following intramuscular delivery of IGF-1. For example, viral mediated expression of IGF-1 in muscle blocks the aging-related loss of skeletal muscle function in mice resulting in increased force production and muscle mass in 27 month old mice injected 9 months earlier (Barron-Davies, 1998). In another study, application of IGF-1 to atrophied old muscles led to substantial increases in muscle mass and recovery of satellite cell activity [151]. Preventing the loss of muscle mass is also achieved by Growth Hormone (GH) administration, associated with increases in IGF-1 levels [152]. Although the use of IGF-1 may have obvious beneficial effects in preventing age-associated muscle loss, serious ethical implications are introduced with respect to its prospective use as a performance-enhancing agent or for cosmetic purposes.

5.   Muscular dystrophies

Muscular Dystrophy refers to a collection of more than 20 inherited disorders affecting approximately 1/3500 live births worldwide [153, 154]. About 2/3 of muscular dystrophies are classified as Duchenne Muscular Dystrophy (DMD) or its milder allelic form, Becker's Muscular Dystrophy caused by loss of function mutations in the X-linked Dystrophin gene. Dystrophin is anchored to the cytoplasmic surface of the myofiber sarcolemma where it forms subsarcolemmal cytoskeletal networks [155]. As such, Dystrophin and its associated proteins (including the dystroglycan complex, and the α-sarcoglycan complex) links the actin-based cytoskeleton of muscle to the extracellular matrix (ECM), thus protecting muscle fibers from contraction induced damage [153, 156]. Dystrophin also interacts with cytoplasmic proteins including members of the Syntrophin family [157, 158]. Syntrophins in turn, are associated with NOS, which mediates intramuscular signaling by releasing NO [158]. Therefore, the Dystrophin complex not only protects myofibers from sarcolemmal damage, but also regulates aspects of muscle function through intracellular signaling cascades.

The contraction induced myofiber damage caused by loss of function of the Dystrophin complex continuously activates resident satellite cells. Repeated cycles of degeneration and regeneration in the skeletal muscle of DMD patients ensues, eventually exhausting the replicative capacity of satellite cells [159-164]. Thus in contrast to muscle pathology in old muscle which reflects inefficient satellite cell activation, the final stages of MD results from reductions in satellite cell replication. This suggests that engraftment of wildtype (dystrophin+) satellite cells in dystrophic muscles would provide an effective therapy for MD.

6.   Cell-based therapies for muscle disease

6.1.   Myoblast transplantation

Cell mediated therapies in which determined myoblasts are introduced into regenerating dystrophic muscle have had limited success [165-170]. The inefficacy of this approach is due largely to the rapid loss of transplanted MPCs from the host muscle immediately after injection as well as the confinement of injected myoblasts to the area of injection. It is estimated that only a small minority (<1%) of transplanted cells survive 4 days following transplantation [171]. Therefore in order to achieve therapeutic benefit, MT would require numerous myoblast injections. These results suggest that efficient MT therapies require the introduction of myogenic cells with "stem cell" properties for repopulation of the satellite cell compartment. In this regard, recent reports describe the isolation of myogenic cells with "stem cell" characteristics [69, 172-175]. An elegant study by Beauchamp et al. [171], describes the characteristics of slowly dividing MPCs with increased capacity for survival following MT. Their results suggest that the ~1% of cells that survive MT may represent the *in vivo* counterparts of reserve cells described in myogenic cell cultures.

   The inability of wildtype myogenic cells to persist in dystrophic muscle may also reflect defects in the ability of *dystrophin-/-* fibers to recruit satellite cells. For example, the Dystrophin associated membrane bound complex may modulate the structure of the sarcolemma thereby interfering with satellite cell recruitment. Although MT still holds promise for clinical application, the biological properties of adult stem cells may hold the key for efficient cell mediated therapy in MD and other degenerative conditions.

6.2.   Stem-cell transplantation

The documented plasticity of adult stem cells isolated from a number of tissues raises the potential for these cells to be isolated and used for the treatment of a variety of degenerative conditions including MD. Given the capacity for pluripotent stem cells to circulate in the bloodstream and home to areas of regeneration, it may be feasible to deliver stem cells intravenously and allow them to infiltrate degenerating target tissues. Stem cells may also have the distinct advantage of providing a long-term source of myogenic progenitors in the case of muscular dystrophy. Stable engraftment of stem cells that give rise to satellite cells would provide a source of myogenic progenitors to repair damaged fibers throughout the patients lifespan.

Several issues of course will need to be resolved before stem cell transplantation can become a clinical reality. It remains to be determined if stem cell transplantations can achieve high degrees of stable engraftment into host tissues. Furthermore, isolating large numbers of stem cells, which maintain their capacity to differentiate, will be essential for therapeutic application. To circumvent some of the difficulties associated with immune rejection, it may be possible to isolate adult stem cells from an affected individual and introduce genetic information *ex vivo* for subsequent transplantation. In the case of muscular dystrophy for instance, adult stem cells from muscle could be isolated and infected with virus containing a functional Dystrophin gene for subsequent transplantation. Inherent in this approach is the requirement for efficient stem cell transduction and *ex vivo* cell expansion, matters that remain to be resolved.

Stem cell therapies may also be improved by identifying growth factors or molecules, which facilitate subsequent differentiation. For example, the introduction of developmental control genes into stem cells may direct their differentiation along a particular pathway. Introduction of Pax7 into stem cells may result in the formation of satellite cell progenitors, which would readily be incorporated into regenerating muscle. Further knowledge concerning molecules involved in stem cell differentiation will likely lead to the advent of effective strategies for treating a variety of degenerative diseases.

7.   Conclusions and perspectives

In recent history, muscle biology has been focused largely on myogenic specification and muscle differentiation in the embryo. A large body of knowledge has been accumulated with regards to the molecular events that control the specification and differentiation of myogenic progenitors in the somite. The study of adult muscle development has not advanced as rapidly, although revitalized interest in adult stem cells recently has led to a number of important and unanticipated discoveries, including the plasticity of muscle stem cells. Although many exciting observations have been made, it is apparent that there remains a large gap in our understanding of the molecular events, which regulate adult stem cell development and differentiation. At present, molecular biologists are presented with the opportunity and challenge of identifying the genetic networks and mechanisms to explain the biological activity of muscle and other tissue-specific stem cells.

For instance, do adult stem cells, which stain poorly with Hoechst dye, have a physiological capacity in organ development and regeneration separable from their ability to be manipulated and injected into animals undergoing regenerative processes? If so, are SP derived stem cells physiologically recruited to form muscle *in vivo*? Pertaining to this question, the possibility that MSCs give rise to satellite cells in a Pax7 dependent fashion requires further experimentation. Where are the MSCs located in adult skeletal muscle tissue? The myogenic progenitors associated with embryonic blood vessels identified by Cossu's group may represent the same population of cells as the MSCs within the SP fraction. In this regard, it will be important to determine if MSCs are present in the embryo and also whether MSCs are associated with blood vessels in adult muscle.

Perhaps one of the most important areas of investigation will be to identify the environmental cues and molecular signals that regulate the plasticity and differentiation of muscle stem cells in response to different conditions. These studies will be essential for optimizing cell-based therapies for the replacement of myogenic progenitors.

The identification and characterization of muscle stem cells in the adult may have profound implications for the future of combined cell and gene therapies in the treatment of a variety of degenerative disorders. By introducing developmental control genes into stem cells isolated from adult tissues, it may be possible to bias their differentiation along a particular lineage. This approach may allow rapid *ex vivo* recovery of specific progenitors with the capacity to repair damaged tissues.

8.   Acknowledgements

We thank Dr. Atsushi Asakura and Dr. Sophie B.P. Chargė for helpful discussions and for careful reading of the manuscript. The work form the laboratory of M.A.R was supported by grants from the National Institutes of Health, the Canadian Institutes of Health Research, the Muscular Dystrophy Association, the Human Frontier's Science Program, and the Canada Research Chair Program. P.S. is supported by a studentship from the Canadian Institutes of Health Research. M.A.R. is a Research Scientist of the Canadian Institutes of Health Research, holds the Canada Research Chair in Molecular Genetics and is a member of the Canadian Genetic Disease Network.

9.   References

1.   Mauro, A. (1961) J. Biophys. Biochem. Cytol. 9: p. 493-495.
2.   Bischoff, R. (1994) A.G. Engel and C. Franszini-Armstrong, Editors. McGraw-Hill: New York. p. 97-118.
3.   Bischoff, R. (1990) Development 109(4): p. 943-52.
4.   Armand, O., et al. (1983) Arch Anat Microsc Morphol Exp, 72(2): p. 163-81.
5.   Cifuentes-Diaz, C., et al. (1993) Neuromuscul Disord, 3(5-6): p. 361-5.
6.   Irintchev, A., et al. (1994)Dev Dyn 199(4): p. 326-37.
7.   Schultz, E. (1976) Am J Anat 147(1): p. 49-70.
8.   Moss, F.P. and Leblond, C.P. (1971)Anat Rec 170(4): p. 421-35.
9.   Schultz, E. (1989) Med Sci Sports Exerc 21(5 Suppl): p. S181-6.
10. Schultz, E. (1996) Dev Biol 175(1): p. 84-94.
11. Gibson, M.C. and Schultz, E. (1983) Muscle Nerve 6(8): p. 574-80.
12. Grounds, M.D. and Yablonka-Reuveni, Z. (1993) Mol Cell Biol Hum Dis Ser 3: p. 210-56.
13. De Angelis, L., et al. (1999) J Cell Biol 147(4): p. 869-78.
14. Cossu, G., et al. (1985) Exp Cell Res 160(2): p. 403-11.
15. Feldman, J.L. and Stockdale, F.E. (1992) Dev Biol 153(2): p. 217-26.
16. Hartley, R.S., Bandman, E. and Yablonka-Reuveni, Z. (1992) Dev Biol 153(2): p. 206-16.
17. Tajbakhsh, S. and Cossu, G. (1997) Curr Opin Genet Dev 7(5): p. 634-41.
18. Cossu, G., et al. (1996) Development 122(2): p. 429-37.
19. Cossu, G., Tajbakhsh, S. and Buckingham, M. (1996) Trends Genet 12(6): p. 218-23.

20. Dietrich, S., *et al.* (1999) Development 126(8): p. 1621-9.
21. Tremblay, P., *et al.* (1998) Dev Biol 203(1): p. 49-61.
22. Dzierzak, E. (1999) Ann N Y Acad Sci 872: p. 256-62; discussion 262-4.
23. Pardanaud, L., *et al.* (1996) Development 122(5): p. 1363-71.
24. Pardanaud, L. and Dieterlen-Lievre, F. (1999) Development 126(4): p. 617-27.
25. Cossu, G. and Mavilio, F. (2000) J Clin Invest 105(12): p. 1669-74.
26. Bianco, P. and Cossu, G. (1999) Exp Cell Res 251(2): p. 257-63.
27. Appell, H.J., Forsberg, S. and Hollmann, W. (1988) Int J Sports Med 9(4): p. 297-9.
28. Darr, K.C. and Schultz, E. (1987) J Appl Physiol 63(5): p. 1816-21.
29. Rosenblatt, J.D., Yong, D. and Parry, D.J. (1994) Muscle Nerve 17(6): p. 608-13.
30. Schultz, E., Jaryszak, D.L. and Valliere, C.R. (1985) Muscle Nerve 8(3): p. 217-22.
31. Hughes, S.M. and Blau, H.M. (1990) Nature 345(6273): p. 350-3.
32. Watt, D.J., *et al.* (1987) Anat Embryol 175(4): p. 527-36.
33. Yablonka-Reuveni, Z. and Rivera, A.J. (1994) Dev Biol 164(2): p. 588-603.
34. Kablar, B., *et al.* (1997) Development 124(23): p. 4729-38.
35. Kablar, B., *et al.* (1998) Biochem Cell Biol 76(6): p. 1079-91.
36. Ordahl, C.P. and Williams, B.A. (1998) Bioessays 20(5): p. 357-62.
37. Megeney, L.A. and Rudnicki, M.A. (1995) Biochem Cell Biol 73(9-10): p. 723-32.
38. Arnold, H.H. and Winter, B. (1998) Curr Opin Genet Dev 8(5): p. 539-44.
39. Cooper, R.N., *et al.* (1999) J Cell Sci 112(Pt 17): p. 2895-901.
40. Smith, C.K., 2nd, Janney, M.J. and Allen, R.E. (1994) J Cell Physiol 159(2): p. 379-85.
41. Beauchamp, J.R., *et al.* (2000) J Cell Biol 151(6): p. 1221-34.
42. Tajbakhsh, S., *et al.* (1996) Dev Dyn 206(3): p. 291-300.
43. Cornelison, D.D. and Wold, B.J. (1997) Dev Biol 191(2): p. 270-83.
44. Rudnicki, M.A. and Jaenisch, R. (1995) Bioessays 17(3): p. 203-9.
45. Rudnicki, M.A., *et al.* (1992) Cell 71(3): p. 383-90.
46. Bulfield, G., *et al.* (1984) Proc Natl Acad Sci U S A 81(4): p. 1189-92.
47. Sicinski, P., *et al.* (1989) Science 244(4912): p. 1578-80.
48. Megeney, L.A., *et al.* (1996) Genes Dev 10(10): p. 1173-83.
49. Cornelison, D.D., *et al.* (2000) Dev Biol 224(2): p. 122-37.
50. Sabourin, L.A., *et al.* (1999) J Cell Biol 144(4): p. 631-43.
51. Yablonka-Reuveni, Z., *et al.* (1999) Dev Biol 210(2): p. 440-55.
52. Braun, T., *et al.* (1992) Cell 71(3): p. 369-82.
53. Kaul, A., *et al.* (2000) Cell 102(1): p. 17-9.
54. Jostes, B., Walther, C. and Gruss, P. (1990) Mech Dev 33(1): p. 27-37.
55. Schafer, B.W., *et al.* (1994) Nucleic Acids Res 22(22): p. 4574-82.
56. Mansouri, A., Goudreau, G. and Gruss, P. (1999) Cancer Res 59(7 Suppl): p. 1707s-1709s; discussion 1709s-1710s.
57. Mansouri, A., *et al.* (1996) Development 122(3): p. 831-8.
58. Seale, P., *et al.* (2000) Cell 102(6): p. 777-86.
59. Tajbakhsh, S., *et al.* (1997) Cell 89(1): p. 127-38.
60. Goulding, M., Lumsden, A. and Paquette, A.J. (1994) Development 120(4): p. 957-71.
61. Bober, E., *et al.* (1994) Development 120(3): p. 603-12.
62. Borycki, A.G. and Emerson, C.P. (1997) Curr Biol 7(10): p. R620-3.
63. Maroto, M., *et al.* (1997) Cell 89(1): p. 139-48.

64. Heanue, T.A., *et al.* (1999) Genes Dev 13(24): p. 3231-43.

65. Relaix, F. and Buckingham, M. (1999) Genes Dev 13(24): p. 3171-8.

66. Seale, P. and Rudnicki, M.A. (2000) Dev Biol 218(2): p. 115-24.

67. Miller, J.B., Schaefer, L. and Dominov, J.A. (1999) Curr Top Dev Biol 43: p. 191-219.

68. Reshef, R., Maroto, M. and Lassar, A.B. (1998) Genes Dev 12(3): p. 290-303.

69. Lee, J.Y., *et al.* (2000) J Cell Biol 150(5): p. 1085-100.

70. Hu, E., Tontonoz, P. and Spiegelman, B.M. (1995) Proc Natl Acad Sci U S A 92(21): p. 9856-60.

71. Teboul, L., *et al.* (1995) J Biol Chem 270(47): p. 28183-7.

72. Schultz, E. and Jaryszak, D.L. (1985) Mech Ageing Dev 30(1): p. 63-72.

73. McGeachie, J.K. and M.D. (1995) Cell Tissue Res 280(2): p. 277-82.

74. Baroffio, A., *et al.* (1996)Differentiation 60(1): p. 47-57.

75. Yoshida, N., *et al.* (1998) J Cell Sci 111(Pt 6): p. 769-79.

76. Carnac, G., *et al.* (2000) Curr Biol 10(9): p. 543-6.

77. Odelberg, S.J., Kollhoff, A. and Keating, M.T. (2000) Cell 103: p. 1099-1109.

78. Allen, R.E., *et al.* (1995) J Cell Physiol 165(2): p. 307-12.

79. Sheehan, S.M., *et al.* (2000) Muscle Nerve 23(2): p. 239-45.

80. Bischoff, R. (1997) Dev Dyn 208(4): p. 505-15.

81. Birchmeier, C. and Gherardi, E. (1998) Trends Cell Biol 8(10): p. 404-10.

82. Tatsumi, R., *et al. (1998)* Dev Biol 194(1): p. 114-28.

83. Schultz, E., *et al.* (1988) Anat Rec 222(1): p. 12-7.

84. Miller, K.J., *et al.* (2000) Am J Physiol Cell Physiol 278(1): p. C174-81.

85. Gal-Levi, R., *et al.* (1998) Biochim Biophys Acta 1402(1): p. 39-51.

86. Leshem, Y., *et al.* (2000) J Cell Physiol 184(1): p. 101-9.

87. Anastasi, S., *et al.* (1997) J Cell Biol 137(5): p. 1057-68.

88. Mason, I.J. (1994) Cell 78(4): p. 547-52.

89. Kuschel, R., Yablonka-Reuveni, Z. and Bornemann, A. (1999) J Histochem Cytochem 47(11): p. 1375-84.

90. Yablonka-Reuveni, Z., Seger, R., and Rivera, A.J. (1999) J Histochem Cytochem 47(1): p. 23-42.

91. Sheehan, S.M. and Allen, R.E. (1999) J Cell Physiol 181(3): p. 499-506.

92. Doumit, M.E., Cook, D.R. and Merkel, R.A. (1993) J Cell Physiol 157(2): p. 326-32.

93. Rando, T.A. and Blau, H.M. (1994) J Cell Biol 125(6): p. 1275-87.

94. Anderson, J.E., *et al.* (1995) Exp Cell Res 216(2): p. 325-34.

95. Mitchell, C.A., McGeachie, J.K. and Grounds, M.D. (1996) Growth Factors 13(1-2): p. 37-55.

96. Floss, T., Arnold, H.H. and Braun, T. (1997) Genes Dev 11(16): p. 2040-51.

97. Fiore, F., Sebille, A. and Birnbaum, D. (2000) Biochem Biophys Res Commun 272(1): p. 138-43.

98. Benito, M., Valverde, A.M. and Lorenzo, M. (1996) Int J Biochem Cell Biol 28(5): p. 499-510.

99. Liu, J.P., *et al.* (1993) Cell 75(1): p. 59-72.

100. Powell-Braxton, L., *et al.* (1993) Genes Dev 7(12B): p. 2609-17.

101. Powell-Braxton, L., *et al.* (1993) Ann N Y Acad Sci 692: p. 300-1.

102. Adams, G.R. and Haddad, F. (1996) J Appl Physiol 81(6): p. 2509-16.

103. Adams, G.R. and McCue, S.A. (1998) J Appl Physiol 84(5): p. 1716-22.

104. Edwall, D., *et al.* (1989) Endocrinology 124(2): p. 820-5.

105. Musaro, A., *et al.* (1999) Nature 400(6744): p. 581-5.

106. Semsarian, C., *et al.* (1999) Nature 400(6744): p. 576-81.

107. Shibasaki, F., *et al.* (1996) Nature 382(6589): p. 370-3.

108. Lawlor, M.A. and Rotwein, P. (2000) J Cell Biol 151(6): p. 1131-40.

109. Lawlor, M.A. and Rotwein, P. (2000) *J*. Mol Cell Biol 20(23): p. 8983-95.

110. Anderson, J.E. (2000) Mol Biol Cell 11(5): p. 1859-74.

111. Orimo, S., *et al.* (1991) Muscle Nerve 14(6): p. 515-20.

112. Tidball, J.G. (1995) Med Sci Sports Exerc 27(7): p. 1022-32.

113. Cantini, M., *et al.* (1994) Biochem Biophys Res Commun 202(3): p. 1688-96.

114. Cantini, M. and Carraro, U. (1995) J Neuropathol Exp Neurol 54(1): p. 121-8.

115. Merly, F., *et al.* (1999) Muscle Nerve 22(6): p. 724-32.

116. Lescaudron, L., *et al.* (1999) Neuromuscul Disord 9(2): p. 72-80.

117. Austin, L., *et al.* (1992) J Neurol Sci 112(1-2): p. 185-91.

118. Kurek, J.B., *et al.* (1996) Muscle Nerve 19(10): p. 1291-301.

119. Kurek, J.B., *et al.* (1997) Muscle Nerve 20(7): p. 815-22.

120. Austin, L., *et al.* (2000) Muscle Nerve 23(11): p. 1700-5.

121. Jesse, T.L., *et al.* (1998) J Cell Biol 140(5): p. 1265-76.

122. Rosen, G.D., *et al.* (1992) Cell 69(7): p. 1107-19.

123. Yang, J.T., *et al.* (1996) J Cell Biol 135(3): p. 829-35.

124. Ferrari, G., *et al.* 1998) *[published erratum appears in Science 1998 Aug 14;281(5379):923]*. Science 279(5356): p. 1528-30.

125. Bittner, R.E., *et al.* (1999) Anatomy and Embryology 199(5): p. 391-6.

126. Gussoni, E., *et al.* (1999) Nature 401(6751): p. 390-4.

127. Goodell, M.A., *et al.* (1996) J Exp Med 183(4): p. 1797-806.

128. Goodell, M.A., *et al.* (1997) Nat Med 3(12): p. 1337-45.

129. Jackson, K.A., Mi, T. and Goodell, M.A. (1999) Proc Natl Acad Sci U S A 96(25): p. 14482-6.

130. Bjornson, C.R., *et al.* (1999) Science 283(5401): p. 534-7.

131. Mezey, E., *et al.* (2000) Science 290(5497): p. 1779-82.

132. Eglitis, M.A. and Mezey, E. (1997) Proc Natl Acad Sci U S A 94(8): p. 4080-5.

133. Alison, M.R., *et al.* (2000) Nature 406(6793): p. 257.

134. Lagasse, E., *et al.* (2000) Nat Med 6(11): p. 1229-34.

135. Pereira, R.F., *et al.* (1998) Proc Natl Acad Sci U S A 95(3): p. 1142-7.

136. Fuchs, E. and Segre, J.A. (2000) Cell 100(1): p. 143-55.

137. Galli, R., *et al.* (2000) Nat Neurosci 3(10): p. 986-91.

138. Rolink, A.G., *et al.* (1999) Nature 401(6753): p. 603-6.

139. Nutt, S.L., *et al.* (1999) Nature 401(6753): p. 556-62.

140. Lamberts, S.W., van den Beld, A.W. and van der Lely, A.J. (1997) Science 278(5337): p. 419-24.

141. Carlson, B.M. (1995) J Gerontol A Biol Sci Med Sci 50 Spec No: p. 96-100.

142. Tzankoff, S.P. and Norris, A.H. (1977) J Appl Physiol 43(6): p. 1001-6.

143. Brooks, S.V. and Faulkner, J.A. (1994) Med Sci Sports Exerc 26(4): p. 432-9.

144. Faulkner, J.A., Brooks, S.V. and Zerba, E. (1995) J Gerontol A Biol Sci Med Sci 50 Spec No: p. 124-9.

145. Johnson, S.E. and Allen, R.E. (1993) J Cell Physiol 154(1): p. 39-43.

146. Dodson, M.V. and Allen, R.E. (1987) Mech Ageing Dev 39(2): p. 121-8.

147. Schultz, E. and Lipton, B.H. (1982) Mech Ageing Dev 20(4): p. 377-83.

148. Decary, S., *et al.* (1997) Hum Gene Ther 8(12): p. 1429-38.

149. Nnodim, J.O. (2000) Mech Ageing Dev 112(2): p. 99-111.

150. Grounds, M.D. (1998) Ann N Y Acad Sci 854: p. 78-91.

151. Chakravarthy, M.V., Davis, B.S. and F.W. (2000) J Appl Physiol 89(4): p. 1365-79.

152. Papadakis, M.A., *et al.* (1996) Ann Intern Med 124(8): p. 708-16.

153. Ozawa, E., *et al.* (1998) Muscle Nerve 21(4): p. 421-38.

154. van Essen, A.J., *et al.* (1992) Hum Genet 88(3): p. 258-66.

155. Bonilla, E., *et al.* (1988) Cell 54(4): p. 447-52.

156. Ervasti, J.M., *et al.* (1990) Nature 345(6273): p. 315-9.

157. Brenman, J.E., *et al.* (1995) Cell 82(5): p. 743-52.

158. Bredt, D.S. (1999) Nat Cell Biol 1(4): p. E89-91.

159. Reimann, J., Irintchev, A. and Wernig, A. (2000) Neuromuscul Disord, 10(4-5): p. 276-82.

160. Ontell, M. (1986) Hum Pathol 17(7): p. 673-82.

161. Ontell, M. (1981) Muscle Nerve 4(3): p. 204-13.

162. Ontell, M., *et al.* (1984) Anat Rec 208(2): p. 159-74.

163. Blau, H.M., Webster, C. and Pavlath, G.K. (1983) Proc Natl Acad Sci U S A 80(15): p. 4856-60.

164. Wright, W.E. (1985) Exp Cell Res 157(2): p. 343-54.

165. Vilquin, J.T., *et al.* (1996) J Cell Biol 133(1): p. 185-97.

166. Partridge, T., *et al.* (1998) Nat Med 4(11): p. 1208-9.

167. Tremblay, J.P., *et al.* (1993) Cell Transplant 2(2): p. 99-112.

168. Gussoni, E., *et al.* (1992) Nature 356(6368): p. 435-8.

169. Blau, H.M. and Springer, M.L. (1995) N Engl J Med 333(23): p. 1554-6.

170. Partridge, T.A. and Davies, K.E. (1995) Br Med Bull 51(1): p. 123-37.

171. Beauchamp, J.R., *et al.* (1999) J Cell Biol 144(6): p. 1113-22.

172. Torrente, Y., *et al.* (2001) J Cell Biol 152(2): p. 335-348.

173. Qu, Z., *et al.* (1998) J Cell Biol 142(5): p. 1257-67.

174. Smith, J. and Schofield, P.N. (1997) Cell Growth Differ 8(8): p. 927-34.

175. Gussoni, E., Blau, H.M. and Kunkel, L.M. (1997) Nat Med 3(9): p. 970-7.

CHAPTER 9

ROBERT L. JILKA and A. MICHAEL PARFITT

Table of contents

1.   Introduction and background

Bone provides the mechanical means of locomotion and protects internal organs. In adults, it consists of a mineralized connective tissue matrix arranged either in a compact or cancellous geometry. The matrix contains embedded osteocytes connected by canaliculi. Most of the bone surface is covered by quiescent lining cells, while surfaces undergoing remodeling are covered by bone resorbing osteoclasts and bone forming osteoblasts. Pluripotent stem cells give rise to the chondroblasts and osteoblasts that form bone during the initial stages of development. Some of these cells, as well as

Stem Cells: A Cellular Fountain of Youth.   Ed. by Mark P. Mattson and Gary Van Zant. 201 — 221

unipotent progeny, remain in the tissue to supply cells needed during subsequent bone growth and modeling, and for the maintenance of the skeleton by remodeling, which occurs throughout life. The differentiated progeny must arise from the stem cells in a coordinated fashion, at the appropriate time and place, and in the variety of hormonal conditions that characterize the prepubertal, pubertal and adult phases of life.

Stem cells have been traditionally associated with epithelia and hematopoiesis in which the cells are of short lifespan and so are in need of continued replacement, with consequent high cell turnover. More recently, stem cells have been found to mediate repair and regeneration in tissues with very long-lived cells such as neurons and myocytes. In bone, stem cells are of the traditional kind, yet bone has its own long-lived cells, osteocytes, which are replaced indirectly. In epithelia, the stem cells are anatomically segregated in a founder cell compartment, but in bone, as in bone marrow, the stem cells lack such a simple geographic means of identification. This circumstance delayed their recognition for many years, but their existence has been inferred from the particular features of bone remodeling and the general features of cell kinetics. Committed osteoclast and osteoblast precursors are derived from pluripotent hematopoietic and stromal stem cells, respectively; but chondroblast precursors are unipotent.

The chapter begins with a description of bone growth and remodeling, and is followed by a discussion of the properties and regulation of the stem cells that give rise to chondroblasts, osteoclasts and osteoblasts. We will then discuss the role of these stem cells in the changes in bone growth or remodeling that occur during adolescence, during aging, and in response to loss of sex steroids at the menopause, or excessive glucocorticoids. Finally, we will address the potential therapeutic utility of osteogenic stem cells.

1.1.   Bone growth and remodeling

The turnover of bone as a structural material must be distinguished from the turnover of its cells. Excluding the marrow, most of the bulk of any bone as an organ consists of a mineralized connective tissue matrix which is hard and rigid, and which is permeated by the osteocyte canalicular network comprising about 2% of the volume [1, 2]. For reasons, that are incompletely understood, bone loses structural integrity as it ages, and must eventually be replaced. The cellular mechanism of replacement, which involves resorption of existing bone by osteoclasts and laying down of new bone in its place by osteoblasts, is referred to as remodeling. Averaged throughout the whole skeleton the rate of replacement is normally about 10%/y, corresponding to a mean lifespan of about 10 years [3].

Bone remodeling is a surface phenomenon so that new bone formation is by apposition. Bones have two principal surfaces: an outer surface that includes the periosteum and articular cartilage, and an inner surface, the endosteum, which is a thin membrane consisting of flat lining cells lying over specialized connective tissue, which encloses the bone marrow [4]. Reflecting the two main structural types of bone, the endosteum has three subdivisions that are in continuity - cancellous, endocortical and intracortical or haversian. Cancellous bone is a three dimensional lattice of plates and rods about 100-150 μm in thickness or diameter, filled in the central skeleton with red marrow, and in the ends of the long bones with yellow marrow. The surface of cancellous bone merges with the inner or endocortical surface of the cortices, which in turn merges with the linings of the haversian canals that traverse the otherwise solid cortices.

During growth most of the periosteum undergoes vigorous and almost continuous bone formation, and most of the endocortical surface undergoes almost equally vigorous and continuous bone resorption. The consequent increase in bone width is matched by proportional increase in length, because of chondroblast proliferation in the growth plate and subsequent replacement of new cartilage by cancellous bone, referred to as endochondral ossification [5] and described further below. Periosteal osteoblasts are derived from local precursor cells, which most likely are stem cells [6], but have been little studied from that standpoint. More is known about the production of osteoblasts in endochondral ossification [7, 8], but the precursors of these cells, which may include vascular pericytes [9], are not well characterized. Periosteal bone formation becomes almost dormant after attainment of skeletal maturity [10], but can be reactivated during fracture healing and in several uncommon diseases. Because of the paucity of information, our own interests, and the general theme of this book, we will say no more about the role of osteoblasts and their precursors in bone growth and repair but will focus on their role in bone remodeling. However, for reasons that will become evident, we will also discuss the status of germinal growth plate chondrocytes as stem cells and the role of chondroblast proliferation in endochondral ossification.

1.2.   Endochondral ossification

At the end of a growing long bone the epiphysis adjacent to the articular cartilage, representing a secondary center of ossification, is separated from the metaphysis and diaphysis by a cartilaginous disc – the growth plate or physis. Its functions are to subserve longitudinal growth, provide the scaffold for deposition of the first metaphyseal cancellous bone by osteoblasts, and ensure correct alignment of new trabeculae for load bearing [11]. All the cells in the growth plate are usually referred to as chondrocytes, but we will refer to cells that make new cartilage as *chondroblasts* and to cells that reside in cartilage but do not make it as *chondrocytes* [12]. Unlike bone, the growth of cartilage is interstitial as chondroblasts make new matrix on all sides rather than only on one side as do osteoblasts. As usually depicted with the epiphysis at the top (Figure 1), the growth plate is organized *vertically* as transverse zones and *horizontally* as vertical columns of cells [5, 12, 13]. At the upper (or distal) end of the growth plate new cartilage is produced, and at the lower (or proximal) end old cartilage is replaced by new bone. For most of the growth period, these processes occur at the same rate, so that the width of the plate does not change. Because the entire complex is growing away from the mid point of the shaft, the different spatially separated zones present at the same *time* represent successive changes occurring in the same *place*.

In the uppermost zone closest to the epiphysis, the germinal zone, also referred to as the resting or reserve zone, the cells are oval to spherical in shape, randomly distributed, widely separated, and lacking columnar arrangement. Cells in the germinal

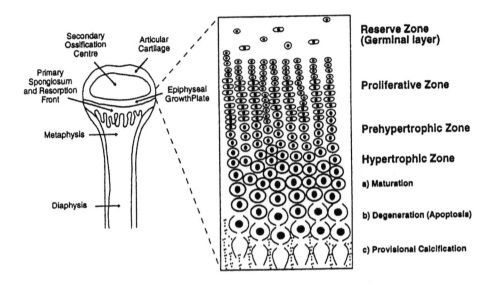

Figure 1.   The epiphyseal growth plate. The right hand panel shows a detailed representation of chondroblasts and chondrocytes in the individual zones of the growth plate. From Stevens and Williams [5], reprinted with permission.

zone adjacent to the upper ends of the cell columns can be considered as stem cells which feed cells into the proliferative zone [12]. Here the cells are more flattened and begin to divide transversely, the daughter cells initially lying side by side before they assume vertical orientation at the onset of intercellular matrix deposition. After completion of proliferation the cells begin to enlarge, forming the hypertrophic zone, and eventually attaining a four fold increase in height, a 50% increase in width, a ten fold increase in volume and a 5 fold reduction in number of cells per unit volume of growth plate [13]. The cells continue to mature, and many eventually die by apoptosis coincident with provisional calcification of the cartilage, and preparatory to vascular invasion from the adjacent metaphysis, followed by resorption of some of the calcified cartilage and deposition of new bone on the surfaces of the cartilage remnants. At the upper tibial growth plate of a 35 day old rat, the heights of the zones are about 50 μm for the germinal zone, about 200 μm for the proliferative zone, and about 350 μm for the hypertrophic zone, for a total of about 600 μm, which corresponds to a duration of about 2 days [12, 13].

1.3.   Bone remodeling and the BMU concept

Bone remodeling refers to the replacement of effete bone material from an internal surface of a bone as an organ by means of successive bone resorption and bone formation. Resorption and formation are commonly regarded as independent processes but in reality they are closely linked within discrete, temporary, anatomic structures first identified by Harold Frost and named by him basic multicellular units, usually abbreviated to BMU [14]. A fully developed BMU (Figure 2) consists of a group of osteoclasts in front, a group of osteoblasts behind, some form of blood supply, often a nerve, and associated connective tissue. The BMU concept has profound implications for bone biology in general but its relevance to this chapter is that it is the behavior of the BMU in time and space which establishes that its executive cells are of comparatively short lifespan (very much shorter than the lifespan of the bone that they reconstruct) and so are in need of continual replenishment.

Most studies of osteoclast and osteoblast stem cells have started with the bone marrow, and most clinical disorders of the aging skeleton involve bone in contact with the bone marrow. Nevertheless, the BMU concept arose from the study of intracortical bone remodeling, and it is here that the unique individuality of the BMU is most obvious (Figure 2). Tetracyclines are fluorescent molecules that are trapped at sites of bone formation, remaining in the same location until the bone is remodeled [3]. Haversian canals run roughly parallel to the long axis of the bone so that an almost complete haversian BMU can be captured in a correctly aligned longitudinal section [15]. Such sections enable the *longitudinal* distance between tetracycline labels to be measured, and such measurements have indicated a rate of longitudinal advance of about 40 μm/d in young dogs and about 25 μm/d in adult humans [14, 15]. The advance continues for about four months during which the BMU maintains the same spatial and temporal relationships between its different components.

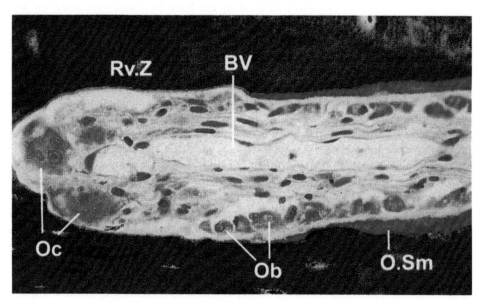

Figure 2.   The BMU. The figure shows the structure traveling from right to left in cortical human bone. Bone resorbing multinucleated osteoclasts (Oc) are followed by osteoblasts (Ob) laying down bone matrix recognized as osteoid (O.sm) prior to mineralization. The reversal zone is in between (Rv.Z). In the center is a blood vessel (BV), either a sinusoid or a large capillary. Original image provided courtesy of Dr. Robert Schenk.

The osteoclasts at the front end of the advancing BMU, forming the so called "cutting cone", are formed by fusion of mononuclear precursors that originate in the bone marrow, circulate in the blood and leave the circulation where they are required. By analogy with extravasating leukocytes, this probably involves an "area code" type mechanism mediated by temporary changes in expression of surface proteins (e.g. integrins) in osteoclast progenitors that recognize selectins in specific endothelial cells [16, 17]. The osteoclasts can be regarded as a team, in the sense of a collection of cells assembled at the same time and place to accomplish a common purpose. The precise 3-dimensional pathway of individual osteoclast movement is unclear, but as the entire BMU moves forward, the osteoclasts collectively resorb laterally for a distance of 100-150 μm away from the central axis. The cells then die by apoptosis and resorption ceases at that cross sectional location [18]; the temporal and causal relationship between these events is controversial. From the distribution of labeled cells at different times after ^{3}H thymidine administration, the average lifespan of an osteoclast nucleus is about 12 days, representing a turnover of about 8%/d [15].

The resorbed bone begins to be replaced soon after the cavity has enlarged laterally to its maximum size. A team of new osteoblasts arrive at the perimeter of the cavity, forming a single ring of cells that lays down new bone centripetally until the cavity has been narrowed to the diameter of a haversian canal, about 40 μm. Based on measurements of the *transverse* distance between tetracycline labels, this takes approximately three months. During this process the osteoblasts become progressively flatter and wider; some become entombed in the bone as osteocytes, some die by apoptosis, and those remaining when the process is completed become the lining cells that cover the new quiescent surface [14, 19]. The lifespan of osteoblasts while they are *making* bone matrix varies from a few to about 100 days, but their lifespan after they have *stopped making* bone matrix and have become osteocytes or lining cells is measured in years. Depending on the rate of longitudinal advance of the BMU a new circular team of osteoblasts arrives about every 8-12 hours, during which the previous team has deposited about 0.3 to 0.5 μm thickness of matrix. The complete set of staggered adjacent teams extends over the closing cone that is about 4-5 times longer than the cutting cone.

The individuality of the BMU is difficult to appreciate in cancellous bone because of its geometric complexity. The BMU cannot travel for long in a straight line and is rarely captured in randomly oriented sections. Nevertheless, there is compelling evidence that bone remodeling is carried out by BMU in cancellous as well as in cortical bone, but instead of excavating a tunnel *through* the bone, they excavate a trench *across* the surface of the bone [14, 18]. Thus, cancellous BMU resemble the bottom half of a bisected cortical BMU [14]. There is the same spatial and temporal relationship between the osteoclast and osteoblast teams that is maintained during BMU advance, the same dependence on the circulation for the supply of osteoclast precursor cells and the same localization of osteoblast assembly to a narrow zone between the cutting and closing "hemicones". The time needed to complete bone formation at each cross sectional location is similar, although the thickness of new bone and the rate of apposition are both smaller. Osteoblasts have the same three fates, although the proportions are different [20].

An important difference between cortical and cancellous BMU is the nature of the blood supply. In the center of a cortical cutting cone is a dilated capillary connected to an arteriole and venule that run through the closing cone and thence to the haversian canal, connecting eventually with vessels in the bone marrow [17]. A cancellous BMU appears to be in direct contact with a vascular sinusoid, beneath a canopy of cells that resemble lining cells and exhibit some phenotypic features of osteoblasts, forming the so called bone remodeling compartment [21]. A final point is the need to distinguish between absolute and relative osteoblast recruitment. The total rate of osteoblast formation per unit of bone surface depends largely on the total number of BMU's, which is related to the rate of bone turnover and so varies substantially between regions and between subjects. Within the BMU, the adequacy of osteoblast supply depends on the magnitude of the task set by osteoclastic resorption. It is for this reason that osteoblast recruitment can be *increased* for the whole skeleton and yet *decreased* relative to local demands [3].

2.   The stem cells of bone

The above discussion makes evident the need for stem cells during skeletal development and growth, as well as the bone remodeling that occurs throughout life. Production of the appropriate number of executive cells at the right time and place during development and remodeling of bone is the result of the orchestration of stem cell self-renewal and differentiation along the appropriate lineage, together with the regulated expansion of early transit amplifying progenitors and subsequent commitment to the differentiated state [22, 23]. At any stage of this process, cells may die by apoptosis - another means of controlling the final number of differentiated cells [24].

Osteoclasts are derived from hematopoietic progenitors of the myeloid lineage, i.e. CFU-GM and CFU-M [25]. Osteoblasts, chondroblasts, as well as the hematopoiesis-supporting stromal cells and adipocytes of the bone marrow, are derived from mesenchymal stem cells [26, 27], also known as marrow stromal fibroblasts. Like their hematopoietic counterparts, the mesenchymal progenitors comprise stem, transit amplifying, and committed compartments, as illustrated for osteoblasts in Figure 3. The defining property of the stem and transit amplifying progenitors is their ability to self-renew. They may divide to produce two identical daughter cells (self-renewal with amplification), or one identical daughter cell and one cell with more differentiated characteristics (self-renewal without amplification), or two cells with a more differentiated phenotype [23]. Stem cells have an extremely high capacity for self-renewal and thus serve as a reservoir of executive cells throughout life. On the other hand, early transit amplifying progenitors are normally proliferating and their self-renewal capacity lasts only for a limited number of cell divisions.

Mesenchymal stem cells are deposited in skeletal tissues during fetal life [26]. Thus, the bone anlage is synthesized by chondroblasts that arise from mesenchymal stem cells, and unipotential chondroblast progenitors arise from these cells to form the growth plate. Mesenchymal stem cells also give rise to the osteoblasts that form the bony collar. The collar is subsequently eroded by osteoclasts that develop from

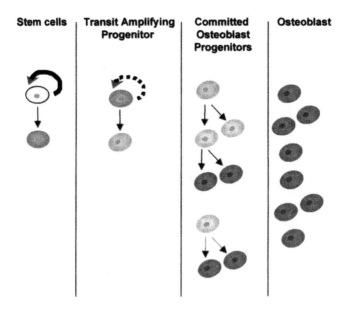

Figure 3.   Osteoblast differentiation. Mesenchymal stem cells (white) with high self-renewal capacity (solid arrow) give rise to early transit amplifying osteoblast progenitors (dark green) with limited self-renewal capacity (dashed arrow). The latter differentiate into late transit amplifying progenitors lacking self-renewal capability (light green). Subsequently, late transit amplifying cells develop into committed osteoblast progenitors (grey) that eventually give rise to fully differentiated osteoblasts (magenta).

circulating progenitors arising in the hematopoietic liver. Subsequent vascular invasion of the marrow cavity brings additional hematopoietic stem cells, as well as mesenchymal stem cells from the periosteum, into the marrow as perivascular cells. The progeny of the mesenchymal stem cells differentiate into osteoblasts, which form the cortical bone, and the primary and secondary spongiosa.

The process of capillary invasion and deposition of osteoclast progenitors from the circulation is seen again during establishment of Haversian remodeling of cortical bone. The source of osteoblast progenitors for cortical bone remodeling is less clear. In view of the fact that the intracortical circulation arises from the endocortical envelope of the marrow, it is possible that osteoblasts derive from pericytes associated with invading capillary endothelial cells [9]. This notion is supported by evidence that pericytes isolated from adult bovine retinal microvessels are capable of forming bone *in vivo* and express phenotypic properties of osteoblasts in culture [28]; and by the close anatomical relationship between blood vessels and sites of bone development [29]. The adventitial reticular cells of the bone marrow stroma might also serve as osteoblast progenitors [30].

2.1.   Osteoclast differentiation

CFU-GM and CFU-M give rise to osteoclasts when given the appropriate signals from stromal support cells [31, 32]. The latter are either closely related to osteoblasts, or are osteoblast precursors; hence they are called stromal/osteoblastic cells. They express on their surface macrophage-colony stimulating factor (M-CSF) and a cytokine called receptor activator of NFkB ligand (RANKL), which act via their respective receptors (c-fms and RANK) on osteoclast progenitors to promote their differentiation [33, 34]. Stromal/osteoblastic cells also secrete osteoprotegerin (OPG) which binds to RANKL, thus acting as a receptor decoy to prevent interaction of RANKL with RANK on osteoclast progenitors. Osteoclast differentiation is subject to both negative and positive control by a complex network of circulating hormones and locally produced cytokines acting on RANKL and OPG synthesis. Parathyroid hormone (PTH) and 1,25-vitamin D_3 are strong inducers of RANKL synthesis as would be expected from their key role in calcium homeostasis. These hormones also decrease the production of OPG. Locally produced cytokines, including interleukin-1 (IL-1), IL-6, IL-11 and tumor necrosis factor (TNF) also directly or indirectly stimulate RANKL synthesis. IL-6 and IL-11 may also influence osteoclastogenesis by stimulating the self-renewal, and inhibiting the apoptosis, of hematopoietic progenitors like CFU-GM [35].

2.2.   Characteristics of mesenchymal stem cells

Besides vascular elements and hematopoietic cells, bone marrow comprises stromal support cells, adipocytes and osteoblasts that are derived from multipotential mesenchymal stem cells [36-38]. These progenitors are also called colony-forming unit-fibroblast (CFU-F) because of their ability to form colonies of fibroblastic cells when placed into tissue culture [36]. Studies with neutralizing antibodies indicate the involvement of PDGF, bFGF, TGFβ and EGF in the initial stages of colony formation [39]. About half of CFU-F colonies contain cells which differentiate into osteoblasts that subsequently form a mineralized matrix when high levels of ascorbic acid are added [40, 41]. Therefore, the subset of CFU-F progenitors that give rise to osteogenic colonies are called CFU-OB.

When transplanted into immunodeficient mice, marrow derived mesenchymal cells form a bony ossicle. The calcified matrix of these ossicles is invaded by vascular elements to form a marrow-like space that supports hematopoiesis and osteoclast formation from host progenitors [41, 42]. This ability to recapitulate the bone and bone marrow development that occurs in fetal life gives confidence that CFU-OB are similar, if not identical, to the stem cells that formed the bony collar during long bone development, and supplied osteoblasts and stromal cells for subsequent development of the bone and bone marrow.

Most preparations of marrow-derived mesenchymal progenitors are heterogeneous when analyzed at the functional level. It was recently reported that human mesenchymal stem cell preparations contain a small population of small cells with distinctive agranular appearance, designated recycling stem cells or (RS-1) cells, as well as small cells with a granular appearance (RS-2) [43]. RS-1 cells appeared to give rise to RS-2 cells, which subsequently proliferate and develop into the large fibroblastic cells that constitute the bulk of adherent mesenchymal progenitors in human marrow cell cultures. The majority of rodent CFU-OB are transit amplifying cells with limited self-renewal capacity [44]; but even the quiescent population does not have the high self-renewal capacity that human progenitors display [45-48]. Culture conditions dramatically affect behavior of mesenchymal progenitors; and in the case of murine progenitors, co-culture with feeder cells is required for optimal colony formation [49].

Despite the above evidence for functional heterogeneity, other studies have indicated that cultures of marrow derived mesenchymal progenitors from human bone marrow, when cultured properly, are homogeneous with regard to expression of specific surface antigens, as well as multipotentiality [50]. Variation in donor population and culture conditions likely account for this discrepancy. Indeed, it has been reported that the proportion of cells expressing the STRO-1 surface antigen varies among donors [51]. Because of the variable heterogeneity of marrow derived mesenchymal osteoblast progenitors, results of studies on the effects of age, hormone status, and other factors on their phenotype must be interpreted with caution. Such heterogeneity makes it very difficult to determine whether, for example, an age-related alteration in the expression of a particular osteoblast phenotype reflects a change in mix of progenitors obtained in the marrow isolate, a change in the program of differentiation of these progenitors, or both.

2.3.   Mesenchymal stem cell differentiation

Although mesenchymal stem cells of the bone marrow can give rise to many cell types, including nerve and muscle cells [52, 53], we will restrict our attention here to formation of chondroblasts, osteoblasts, and the stromal and adipocytic cells of the bone marrow.

Chondroblast differentiation involves a complex feedback loop involving locally produced PTHrP and Indian hedgehog (Ihh), which precisely controls chondrocyte proliferation and differentiation [54, 55]. Ihh is produced by prehypertrophic chondrocytes, and it indirectly stimulates the secretion of PTHrP by periarticular cells. PTHrP then acts on uncommitted prehypertrophic cells to preserve their proliferative state and inhibit differentiation. Thus, when the prehypertrophic chondrocyte population is diminished as they become hypertrophic and die, Ihh levels and thereby PTHrP levels, decrease. This decline results in decreased chondroblast proliferation and increased differentiation of additional prehypertrophic chondrocytes. Other factors such as transforming growth factor-β (TGFβ) and bone morphogenetic proteins (BMPs), and BMP antagonists like noggin, are also involved in the regulation of chondroblast differentiation [56, 57].

Osteoblast differentiation also requires the participation of locally produced growth factors including BMPs, insulin-like growth factors (IGFs), and TGFβ [57, 58], and interaction of progenitors with the extracellular collagenous matrix is also involved [59]. Recent evidence indicates that activin, BMP-2 and BMP-4 are produced endogenously in murine bone marrow cultures, and are required for full expression of the osteoblast phenotype [60, 61]. At the same time, BMPs stimulate the synthesis of the BMP antagonists noggin and gremlin [62, 63], which may serve as part of a negative feedback loop to precisely control the supply of osteoblasts from local CFU-OB.

Stromal cells capable of supporting hematopoiesis develop from mesenchymal stem cell progenitors in cultures of human marrow without the need for additional differentiation factors. This is indicated by their ability to support the hematopoietic capacity of CD34+ hematopoietic stem cells and the secretion of several colony-stimulating factors, interleukins and cytokines [64]. The generation of the stromal/osteoblastic cells that support osteoclast differentiation, on the other hand, appears to depend on the development of osteoblast-like precursors. Indeed, when osteoblast differentiation is prevented either by addition of noggin or by genetic deletion of Osf2/Cbfa1, support of osteoclast differentiation is sharply diminished, as is also the case for cultures from mice with reduced osteogenic capacity [60, 65-67].

The differentiation of adipocytes from mesenchymal progenitors is stimulated by glucocorticoids, as well as by ligands such as thiazolidinediones that activate PPARγ2, a transcription factor required for fat cell development. Glucocorticoids promote this process by stimulating the synthesis of the transcription factor cEBPα, which then activates the promoter for PPARγ2 [68]. Progeny of mesenchymal progenitors also exhibit a remarkable capacity to convert between the osteoblast and adipocyte phenotype even after lipid accumulation has begun and after expression of genes characteristic of the adipocyte phenotype, such as lipoprotein lipase, has been initiated [69-72]. Specific transcription factors orchestrate lineage commitment in these progenitors as evidenced by the finding that activation of PPARγ2 in a marrow-derived osteoblastic cell line stimulated adipogenesis and simultaneously suppressed Osf2/Cbfa1 and the biosynthesis of matrix proteins [73]. Subtypes of the BMP receptor family may also differentially regulate adipogenesis and osteoblastogenesis [74].

2.4. Coordination of osteoclast and osteoblast production for bone remodeling

Both osteoclasts and osteoblasts are needed simultaneously during BMU progression, and bone formation must only occur at sites of previous resorption. Recruitment of osteoblasts may be due to release of growth factors from the bone matrix during bone resorption, which then activate osteoblast differentiation [75]. *In vitro* and *in vivo* evidence indicates that active TGFβ is indeed released from the bone matrix during osteoclastic bone resorption [76, 77]; but despite the well established stimulatory effect of TGFβ on bone formation, its involvement in recruitment of osteoblasts to the BMU has not been demonstrated.

A second potential mechanism is based on the strong association of vascular elements with sites of bone remodeling. The new endothelial cells that arise at the tip of the capillary during the course of BMU angiogenesis may transiently express the "area code" needed for osteoclast progenitors to exit the circulation near the cutting cone. As the capillary advances, the endothelial cells formerly at the tip find themselves adjacent to the resorbed bone where osteoblasts are needed. Consequently, the osteoclast area code is turned off and signals for the proliferation and differentiation of pericytes, or other local osteoblast progenitors, is turned on [9].

The ability of mesenchymal stem cells to differentiate into stromal cells as well as osteoblasts provides the basis of a third mechanism to link osteoclast formation to osteoblast formation [78, 79]. As mentioned above, cells of the osteoblast lineage provide stromal support for osteoclast formation via synthesis of RANKL in response to hormones such as PTH. Mice lacking the Cbfa1 transcription factor needed for osteoblast differentiation, not only lack bone, but also osteoclasts [66]. Moreover, marrow cultures from mice with reduced osteoblast CFU-OB progenitors also exhibit reduced osteoclastogenic capacity [67]. It is unknown whether the cells that support osteoclast formation represent a distinct lineage, or whether they are early osteoblast progenitors that subsequently lose osteoclastogenic support capacity as they differentiate into a matrix-synthesizing cell.

3.   Stem cell regulation of bone growth and remodeling

A variety of circumstances modulate skeletal growth and remodeling. The onset of sex steroid production at puberty leads to closure of the growth plate and reduction in the rate of bone remodeling [80]. The menopause in women, and chemical or surgical castration in women or men increases bone remodeling [81]. Aging is often associated with a reduction in bone remodeling and bone loss [82, 83]. And, administration of high levels of glucocorticoids suppresses bone remodeling and leads to bone loss [84]. Altered production of bone cells from their respective stem cell progenitors is at least partially responsible for each of these changes.

3.1.   Growth plate closure

Chondroblast proliferation and longitudinal bone growth is dependent on growth hormone [85, 86]. Studies of cell kinetics in the growth plate have demonstrated that cell replication in the proliferative zone of the growth plate begins to decline at the end of the pubertal growth spurt, resulting in a reduction in the longitudinal growth of long bones [85, 87]). When proliferation reaches zero, new chondroblasts needed for growth are no longer produced, and both the addition of new cartilage at the distal end of the growth plate and the replacement of old cartilage by bone at the proximal end cease. Instead, thin layers of bone are deposited on both sides of the growth plate upon which additional mineralization proceeds; and this is followed by osteoclastic perforation and eventual removal of this hard tissue barrier [88]. This process of active epiphyseal fusion is mediated by estrogen in both men and women as indicated by failure of growth plate closure and continued longitudinal growth in an a man lacking the α form of the estrogen receptor (ERα) [89] and in aromatase deficient men and women [90]. The mechanism by which estrogen slows, and eventually stops, proliferation of chondroblast progenitors is unknown. Although estrogen could cause early senescence of the replicating cell, loss of estrogen (by ovariectomy) in the growing rat increased the number of proliferating chondroblasts in the proliferative zone [91], suggesting that estrogen does not exert irreversible senescence. As will be discussed further below, estrogen also suppresses CFU-OB replication [48]. It is thus tempting to suggest that the steroid has the same effect on chondroblast stem cell replication.

3.2.   Sex steroids and bone remodeling

Bone remodeling slows during puberty with the onset of sex steroid production. Conversely, bone remodeling increases when estrogen levels decline at the menopause, or when sex steroid production is lost because of other reasons [20]. The increased remodeling reflects the birth of new BMUs or prolongation of the lifespan of existing ones. The bone loss that accompanies this increased remodeling is due to an imbalance between bone formation and resorption in the favor of the latter, perhaps due to loss of pro-apoptotic effects of estrogen and androgen on osteoclasts together with loss of anti-apoptotic effects of these steroids on osteoblasts [24, 81, 92].

Increased production or longevity of BMUs requires increased production of osteoclasts and osteoblasts. Indeed, studies in rodents have demonstrated that loss of sex steroids is accompanied by an increase in the number of CFU-GM osteoclast progenitors and CFU-OB osteoblast progenitors [93-95]. The increase in CFU-GM is due at least in part to loss of suppressive effects of estrogens on the production of cytokines like IL-6 and TNF that stimulate CFU-GM self-renewal and inhibit their apoptosis [35, 81, 96]. The increased cytokine production also leads to increased osteoclast formation. The increased osteoblastogenesis caused by estrogen deficiency results from the removal of inhibitory control of CFU-OB self-renewal by β-estradiol [48]. It is unknown whether this is due to a direct action of estrogens on CFU-OB, or whether, as in the case of CFU-GM, estrogen deficiency unleashes the production of factors that are normally suppressed by the steroid, to cause increased CFU-OB self renewal. Decreased apoptosis may also contribute, but this seems unlikely in the case of CFU-OB since estrogens exert anti-apoptotic effects on cells of the osteoblastic lineage [92].

Because the stromal/osteoblastic cells that support osteoclast development are also derived from early mesenchymal progenitors [97, 98], an increase in CFU-OB self renewal may also contribute to the increase in osteoclast formation. Indeed, ovariectomy failed to increase osteoclastogenesis and bone remodeling in SAMP6 mice [67] that have decreased CFU-OB number. Thus, an increase in the number of cells that support osteoclast development, together with increased production of osteoclastogenic cytokines by these same cells, may be responsible for the increase in osteoclast formation in estrogen deficiency.

3.3.   Glucocorticoid excess

Bone loss occurs in patients taking glucocorticoids to suppress immune cell function in arthritis or following organ transplantation lose bone. The histological features of this condition are decreased bone formation rate, decreased wall thickness, and *in situ* death of portions of bone. Some of these effects are caused by glucocorticoid-induced apoptosis of osteoblasts and osteocytes [99, 100]. The decreased remodeling is most likely due to the reduction in the number of CFU-OB and CFU-GM progenitors that also occurs with excessive glucoroticoids, perhaps via a decrease the production and/or activity of cytokines like IGFs [101] and IL-6 type cytokines [102] that promote the replication of these early progenitors.

3.4.   Age related bone loss

Aging has a negative impact on most regenerating tissues, including bone. The development of age-related osteopenia is due to a decline in the number of osteoblasts present in the BMU, compared to the demand for them. This contention is based on evidence that bone from elderly humans has decreased mean wall thickness — a histologic index of the amount of bone made by a team of osteoblasts [20]. Mean wall thickness is a function of the number and activity of osteoblasts recruited to the remodeling site and the average volume of bone matrix synthesized by each osteoblast. Histomorphometric evidence indicates that the amount of bone formed per unit of osteoblast surface is similar in elderly osteoporotic and non-osteoporotic individuals [20]. Thus, a reduction in osteoblast density, rather than activity, must account for most of the decline in bone formation rate seen with advancing age.

Age-related bone loss may be due in part to the failure of CFU-OB to supply sufficient osteoblasts to the BMU. Numerous studies have shown that aging has a negative impact on the number of early osteoblast progenitors in both animals and humans [67, 83, 103-109]. Recent studies of ours suggest that the decrease in the number of CFU-OB progenitors during aging is due to diminished CFU-OB self-renewal capacity. Thus, self-renewal of CFU-OB from SAMP6 mice, a mouse model of early senescence, is reduced compared to the control SAMR1 mouse strain, and correlates with a decrease in CFU-OB number, osteoblast number and bone formation rate [44, 67]. Preliminary studies have also shown that self-renewal of CFU-OB from C57Bl/6 mice declines with age (R.L. Jilka, unpublished).

Beside this change in proliferative capacity, the fate of early mesenchymal progenitors may be altered during aging. It has long been appreciated that marrow fat increases and osteoblast number decreases with advancing age. In view of the common ancestry of marrow adipocytes and osteoblasts, as well as to the "plastic" nature of more committed progeny, it is possible that bone marrow adipocytes develop at the expense of osteoblasts during aging, thus leading to decreased bone formation and bone loss [73, 109-112]. A clue to the molecular basis of this phenomenon is evidence that activation of the proadipogenic transcription factor PPARγ2 in osteoblastic cells stimulates adipogenesis and irreversibly blocks osteoblast differentiation [73]. Thus, an increase in the expression and/or activation of PPARγ2 might contribute to the development of age-related osteopenia [73]. Consistent with the latter, PPARγ2 ligands derived from fatty acid oxidation products increase with advancing age [113-116], and preliminary studies indicate that some of these products inhibit osteoblast differentiation *in vitro* [117].

4.   Therapeutic potential of mesenchymal progenitors

Because marrow-derived mesenchymal stem cells are multipotential and can be greatly expanded in ex-vivo cultures, they would seem ideal for increasing bone formation in states of osteopenia associated with diminished osteoblast progenitors, as well as for correcting genetic disorders of bone. Several studies have demonstrated the feasibility of this approach using endogenous or retroviral markers to identify and quantify transplanted cells following infusion. Using a canine model of hematopoietic reconstitution involving lethal irradiation and rescue with G-CSF-primed hematopoietic precursors, it was found that most of the mesenchymal stem cells that are also present in these isolates localized to bone marrow [118]. PCR-based quantification indicated that between 0.004% and 0.32% of marrow cells were of donor origin, with the frequency of CFU-F of 0.01%. Moreover, these progenitors persisted in the marrow for as long as 6 months. In another study, human marrow derived mesenchymal progenitors were injected into the peritoneal cavity of fetal sheep [119]. (Rejection of transplanted cells does not occur in this system because of the stage of development, and the similarity of the human and ovine immune systems.) Transplanted cells were detected by PCR analysis of human β2-microglobulin in several tissues including bone marrow and cartilage up to 13 months after transplantation. Engraftment and site specific differentiation of the transplanted progenitors into myocytes, adipocytes, marrow myocytes, marrow and thymic stromal cells and chondrocytes occured as evidenced by immunocytochemical analysis of human β2-microglobulin expression, but human osteoblasts and osteocytes were not examined.

Studies in mice have demonstrated differentiation of transplanted mesenchymal progenitors into marrow stromal cells, osteoblasts and osteocytes [120-123]. In the latter study, mice were transplanted at 6 weeks of age and the tissues analyzed at 6 months of age. Marrow chimerism was greater than 20%, and 0.3% of osteocytes were donor derived. However, the percentage of donor derived osteoblasts could not be determined owing to the extensive growth and remodeling that occurs in murine long bones between 6 weeks and 6 months of age.

Although the above findings are encouraging, successful treatment of genetic diseases of bone with transplanted mesenchymal progenitors requires that the normal donor cells have an advantage over the abnormal progenitors of the recipient, either by reason of overwhelming numbers or proliferative capacity. The problem is illustrated by recent attempts to cure osteogenesis imperfecta (OI), a genetic disorder resulting in generalized osteopenia, bony deformities and fractures, and short stature caused by mutations in the Col1A1 or Col1A2 gene. In a murine model of OI [124], cultured marrow-derived progenitors were infused into 3 week old irradiated mutant mice. Donor cells were found in marrow, spleen, bone, lung, cartilage, and brain, but only modest changes in collagen and mineral content were seen in the following 4 to 10 weeks. Analysis of cultures subsequently established from bone marrow, calvaria, cartilage, dermal tissue skin of transplanted mice indicates that 4-15% were of donor origin.

Transplantation of total allogeneic bone marrow, after ablative conditioning, to three children (aged 13 to 17 months) with OI resulted in engraftment of mesenchymal as well as hematopoietic cells [125, 126]. However, only 1.5 to 2.0% of marrow-derived mesenchymal progenitors were donor derived. Preliminary histologic examination of bone tissue taken 216 days after transplantation indicated improvement of bone formation and mineralization. Moreover, the rate of growth and gain in bone mineral content was superior to two untreated subjects, and approached normal. However, these findings are difficult to reconcile with the small percentage of donor cells in the recipients. It is possible that a putative ablation-induced osteogenic response [127] could have contributed to these findings, but more studies are needed.

5.   Conclusion

Just as stem cells of other self-renewing tissues govern growth and turnover, the behavior of the stem cells that give rise to chondroblasts, osteoclasts and osteoblasts play a dominant role in both longitudinal bone growth and bone remodeling. The epiphyseal closure and reduced bone remodeling that characterize the end of the adolescent growth spurt are consequences of the suppressive effects of sex steroids on the replication of chondroblasts, osteoclasts and osteoblasts from their respective transit amplifying progenitors. On the other hand, the increased bone remodeling due to sex steroid deficiency results from loss of these suppressive effects on progenitors of osteoclasts and osteoblasts. Age-related osteopenia may be due to reduced osteoblast production because of decreased self-renewal of transit-amplifying progenitors and/or diversion of progenitors to the adipocyte instead of osteoblast lineage. Advances in the isolation and transplantation of mesenchymal stem cells offers hope that they can be used to ameliorate osteopenia and bone disease, but this work is in its infancy.

6.   Acknowledgments

The authors thank their colleagues Drs. Stavros Manolagas, Robert S. Weinstein, Teresita Bellido, Charles O'Brien, and Gina DiGregorio-Taguchi for discussions and research contributions leading to the development of the concepts described in this review; and acknowledge the research support of the NIH (P01-AG13918) and the Department of Veterans Affairs (Merit Review and Research Enhancement Awards).

7.   References

1. Jee, W.S.S. (1988) in: L. Weiss (Ed.), Cell and tissue biology, 6, Chapter 7, The skeletal tissues, Urban and Schwarzenberg, Baltimore, pp. 213-254.
2. Baud, C.A. (1973) in: Z.F.G. Jaworski, S. Klosevych and E. Cameron (Eds.), Proceeding of the first workshop on bone morphometry, Histophysiology of the osteocyte: an introduction to the morphometry of per-osteocytic lacunae, University of Ottawa, pp. 267-272.
3. Parfitt, A.M. (1988) in: B.L. Riggs and L.J. Melton (Eds.), Osteoporosis — Etiology, diagnosis and management, Bone remodeling: relationship to the amount and structure of bone and the pathogenesis and prevention of fractures, Raven Press, New York, pp. 45-94.
4. Parfitt, A.M. (1989) in: M. Kleerekoper and S. Krane (Eds.), Clinical disorders of mineral metabolism, Surface specific bone remodeling in health and disease, Mary Ann Liebert Publishers, Inc., New York, pp. 7-14.
5. Stevens, D.A. and Williams, G.R. (1999) Mol Cell Endocrinol 151, 195-204.
6. Owen, M. (1971) in: G.H. Bourne (Ed.), Volume III. Development and Growth, 2nd, Cellular dynamics of bone, Academic Press, New York, NY, pp. 271-298.
7. Kimmel, D.B. and Jee, W.S. (1980) Calcif.Tissue Int. 32, 123-133.
8. Jaworski, Z.F.G., Kimmel, D.B., and Jee, W.S.S. (1983) in: R.R. Recker (Ed.), Bone Histomorphometry: Techniques and Interpretation, Cell kinetics underlying skeletal growth and bone tissue turnover, CRC

Press, Boca Raton, FL, pp. 225-239.

9.   Parfitt, A.M. (2000) Bone 26, 319-323.

10. Balena, R., Shih, M.-S., and Parfitt, A.M. (1992) J. Bone Miner.Res. 7, 1475-1482.

11. Frost, H.M. (1972) The physiology of cartilaginous, fibrous, and boney tissue. Charles C. Thomas, Springfield, IL.

12. Sissons, H.A. (1971) in: G.H. Bourne (Ed.), Volume III. Development and Growth, 2nd, The growth of bone, Academic Press, New York, NY, pp. 148-180.

13. Schenk, R.K. and Hunziker, E.B. (1991) in: F.H. Glorieux (Ed.), Rickets, Growth plate: histophysiology, cell and matrix turnover, New York, NY, pp. 63-78.

14. Parfitt, A.M. (1994) J. Cell. Biochem. 55, 273-286.

15. Jaworski, Z.F.G. (1992) in: B.K. Hall (Ed.), Bone: Bone Metabolism and Mineralization, Vol. 4, Haversian systems and haversian bone, CRC Press, Boca Raton, FL, pp. 21-45.

16. Teitelbaum, S.L., Tondravi, M.M., and Ross, F.P. (1996) in: R.Marcus, D. Feldman and J. Kelsey (Eds.), Osteoporosis, Osteoclast biology, Academic Press, San Diego, CA, pp. 64-94.

17. Parfitt, A.M. (1998) Bone 23, 491-494.

18. Parfitt, A.M., Mundy, G.R., Roodman, G.D., Hughes, D.E., and Boyce, B.F. (1996) J. Bone Miner. Res. 11, 150-159.

19. Jilka, R.L., Weinstein, R.S., Bellido, T., Parfitt, A.M., and Manolagas, S.C. (1998) J. Bone Miner. Res. 13, 793-802.

20. Parfitt, A.M. (1990) in: B.K. Hall (Ed.), Bone. Volume 1. The Osteoblast and Osteocyte, Bone-forming cells in clinical conditions, Telford Press and CRC Press, Boca Raton, FL, pp. 351-429.

21. Hauge, E.M., Qvesel, D., Eriksen, E.F., Mosekilde, L., and Melsen, F. (2001) J. Bone Miner. Res. 16, 1575-1585.

22. Morrison, S.J., Shah, N.M., and Anderson, D.J. (1997) Cell 88, 287-298.

23. Loeffler, M. and Potten, C.S. (1997) in: C.S. Potten (Ed.), Stem Cells, Chapter 1, Stem cells and cellular pedigrees - a conceptual introduction, Academic Press, San Diego, CA, pp. 1-27.

24. Boyce, B.F., Xing, L., Jilka, R.L., Bellido, T., Weinstein, R.S., Parfitt, A.M., and Manolagas, S.C. (2002) in: J.P. Bilezikian, L.G. Raisz and G. Rodan (Eds.), Prinicples of Bone Biology, 2nd Edition Apoptosis in bone cells, Academic Press, San Diege, pp. 151-168.

25. Roodman, G.D. (1999) Exp. Hematol. 27, 1229-1241.

26. Bianco, P., Riminucci, M., Kuznetsov, S., and Robey, P.G. (1999) Crit Rev. Eukaryot. Gene Expr. 9, 159-173.

27. Bianco, P. and Gehron, R.P. (2000) J.Clin.Invest 105, 1663-1668.

28. Doherty, M.J., Ashton, B.A., Walsh, S., Beresford, J.N., Grant, M.E., and Canfield, A.E. (1998) J. Bone Miner. Res. 13, 828-838.

29. M.Brookes (1971) in: The Blood Supply of Bone, Chapter 7, Cortex and Periosteum, Appleton-Century-Crofts, New York, pp. 92-122.

30. Bianco, P. and Riminucci, M. (1998) in: J. Beresford and M.E. Owen (Eds.), Marrow stromal cell culture, Chapter 2, The bone marrow stroma *in vivo*: ontogeny, structure, cellular composition and changes in disease, Cambridge University Press, New York, pp. 11-25.

31. Hattersley, G., Kerby, J.A., and Chambers, T.J. (1991) Endocrinology 128, 259-262.

32. Yamazaki, H., Kunisada, T., Yamane, T., and Hayashi, S.I. (2001) Exp. Hematol. 29, 68-76.

33. Lacey, D.L., Timms, E., Tan, H.L., Kelley, M.J., Dunstan, C.R., Burgess, T., Elliott, R., Colombero, A., Elliott, G., Scully, S., Hsu, H., Sullivan, J., Hawkins, N., Davy, E., Capparelli, C., Eli, A., Qian, Y.X., Kaufman, S., Sarosi, I., Shalhoub, V., Senaldi, G., Guo, J., Delaney, J. and Boyle, W.J. (1998) Cell

93, 165-176.

34. Hofbauer, L.C., Khosla, S., Dunstan, C.R., Lacey, D.L., Boyle, W.J., and Riggs, B.L. (2000) J. Bone Miner. Res. 15, 2-12.

35. Jilka, R.L. (1998) Bone 23, 75-81.

36. Owen, M. (1988) J. Cell Sci. Suppl. 10, 63-76.

37. Aubin, J.E. (1998) Journal of Cellular Biochemistry Supplements 30/31, 73-82.

38. Triffitt, J.T. (1996) in: J.P. Bilezikian, L.G. Raisz and G.A. Rodan (Eds.), Principles of Bone Biology, Chapter 4, The stem cell of the osteoblast, Academic Press, San Diego, pp. 39-50.

39. Kuznetsov, S.A., Friedenstein, A.J. and Robey, P.G. (1997) British Journal of Haematology 97, 561-570.

40. Beresford, J.N., Graves, S.E. and Smoothy, C.A. (1993) Am.J.Med.Genet. 45, 163-178.

41. Kuznetsov, S.A., Krebsbach, P.H., Satomura, K., Kerr, J., Riminucci, M., Benayahu, D. and Robey, P.G. (1997) J. Bone Miner. Res. 12, 1335-1347.

42. Krebsbach, P.H., Kuznetsov, S.A., Satomura, K., Emmons, R.V.B., Rowe, D.W. and Robey, P.G. (1997) Transplantation 63, 1059-1069.

43. Colter, D.C., Class, R., DiGirolamo, C.M. and Prockop, D.J. (2000) Proc. Natl. Acad. Sci. USA 97, 3213-3218.

44. DiGregorio-Taguchi, G.B., Gubrij, I., Smith, C., Parfitt, A.M., Manolagas, S.C. and Jilka, R.L. (2001) J. Bone Miner. Res. 15, S376. (Abstract)

45. Bellows, C.G., Heersche, J.N. and Aubin, J.E. (1990) Dev. Biol. 140, 132-138.

46. Van Vlasselaer, P., Falla, N., Snoeck, H. and Mathieu, E. (1994) Blood 84, 753-763.

47. Falla, N., Van Vlasselaer, P., Bierkens, J., Borremans, B., Schoeters, G. and Van Gorp, U. (1993) Blood 82, 3580-3591.

48. Di Gregorio, G.B., Yamamoto, M., Ali, A.A., Abe, E., Roberson, P., Manolagas, S.C. and Jilka, R.L. (2001) J. Clin. Invest. 107, 803-812.

49. Kuznetsov, S. and Robey, P.G. (1996) Calcif. Tissue Int. 59, 265-270.

50. Pittenger, M.F., Mackay, A.M., Beck, S.C., Jaiswal, R.K., Douglas, R., Mosca, J.D., Moorman, M.A., Simonetti, D.W., Craig, S. and Marshak, D.R. (1999) Science 284, 143-147.

51. Stewart, K., Walsh, S., Screen, J., Jefferiss, C.M., Chainey, J., Jordan, G.R. and Beresford, J.N. (1999) J. Bone Miner. Res. 14, 1345-1356.

52. Mezey, E., Chandross, K.J., Harta, G., Maki, R.A. and McKercher, S.R. (2000) Science 290, 1779-1782.

53. Wang, J.S., Shum-Tim, D., Galipeau, J., Chedrawy, E., Eliopoulos, N. and Chiu, R.C. (2000) J. Thorac. Cardiovasc. Surg. 120, 999-1006.

54. Lanske, B., Karaplis, A.C., Lee, K., Luz, A., Vortkamp, A., Pirro, A., Karperien, M., Defize, L.H.K., Ho, C., Mulligan, R.C., Abou-Samra, A.B., Jüppner, H., Segre, G.V. and Kronenberg, H.M. (1996) Science 273, 663-666.

55. Vortkamp, A., Lee, K., Lanske, B., Segre. G.V., Kronenberg, H.M. and Tabin, C.J. (1996) Science 273, 613-622.

56. Brunet, L.J., McMahon, J.A., McMahon, A.P. and Harland, R.M. (1998) Science 280, 1455-1457.

57. Hogan, B.L. (1996) Genes Dev. 10, 1580-1594.

58. Jilka, R.L. and Manolagas, S.C. (1994) in: R.Marcus (Ed.), Osteoporosis, Chapter 2, The cellular and biochemical basis of bone remodeling, Blackwell Scientific Publications, Cambridge, MA, pp. 17-48.

59. Xiao, G., Wang, D., Benson, M.D., Karsenty, G. and Franceschi, R.T. (1998) J. Biol. Chem. 273, 32988-32994.

60. Abe, E., Yamamoto, M., Taguchi, Y., Lecka-Czernik, B., O'Brien, C.A., Economides, A.N., Stahl, N.,

Jilka, R.L. and Manolagas, S.C. (2000) J. Bone. Miner. Res. 15, 663-673.

61. Gaddy-Kurten, D., Coker, J.K., Abe, E., Jilka, R.L. and Manolagas, S.C. (2001) (submitted)
62. Pereira, R.C., Economides, A.N. and Canalis, E. (2000) Endocrinology 141, 4558-4563.
63. Gazzerro, E., Gangji, V. and Canalis, E. (1998) J. Clin. Invest. 102, 2106-2114.
64. Majumdar, M.K., Thiede, M.A., Mosca, J.D., Moorman, M. and Gerson, S.L. (1998) J. Cell. Physiol. 176, 57-66.
65. Gao, Y.H., Shinki, T., Yuasa, T., Kataoka-Enomoto, H., Komori, T., Suda, T. and Yamaguchi, A. (1998) Biochem. Biophys. Res. Comm. 252, 697-702.
66. Komori, T., Yagi, H., Nomura, S., Yamaguchi, A., Sasaki, K., Deguchi, K., Shimizu, Y., Bronson, R.T., Gao, Y.-H., Inada, M., Sato, M., Okamoto, T., Kitamura, Y., Yoshiki, S. and Kishimoto, T. (1997) Cell 89, 755-764.
67. Jilka, R.L., Weinstein, R.S., Takahashi, K., Parfitt, A.M. and Manolagas, S.C. (1996) J. Clin. Invest. 97, 1732-1740.
68. Rosen, E.D. and Spiegelman, B.M. (2000) Annu. Rev. Cell Dev. Biol. 16, 145-171.
69. Park, S.R., Oreffo, R.O. and Triffitt, J.T. (1999) Bone 24, 549-554.
70. Diascro, D.D.J., Vogel, R.L., Johnson, T.E., Witherup, K.M., Pitzenberger, S.M., Rutledge, S.J., Prescott, D.J., Rodan, G.A. and Schmidt, A. (1998) J. Bone Miner. Res. 13, 96-106.
71. Houghton, A., Oyajobi, B.O., Foster, G.A., Russell, R.G. and Stringer, B.M.J. (1998) Bone 22, 7-16.
72. Nuttall, M.E., Patton, A.J., Olivera, D.L., Nadeau, D.P. and Gowen, M. (1998) J. Bone Miner. Res. 13, 371-382.
73. Lecka-Czernik, B., Gubrij, I., Moerman, E.J., Kajkenova, O., Lipschitz, D.A., Manolagas, S.C. and Jilka, R.L. (1999) J. Cell Biochem. 74, 357-371.
74. Chen, D., Ji, X., Harris, M.A., Feng, J.Q., Karsenty, G., Celeste, A.J., Rosen, V., Mundy, G.R. and Harris, S.E. (1998) J. Cell Biol. 142, 295-305.
75. Mundy, G.R. (1999) in: Bone remodeling and its disorders, Chapter 1, Bone remodeling, Martin Dunitz, Ltd, Malden, MA, pp. 1-11.
76. Pfeilschifter, J. and Mundy, G.R. (1987) Proc. Natl. Acad. Sci. USA 84, 2024-2028.
77. Chirgwin, J.M. and Guise, T.A. (2000) Crit Rev. Eukaryot. Gene Expr. 10, 159-178.
78. Rodan, G.A. and Martin, T.J. (1981) Calcif.Tissue Int. 33, 349-351.
79. Manolagas, S.C., Jilka, R.L., Bellido, T., O'Brien, C.A. and Parfitt, A.M. (1996) in: J.P. Bilezikian, L.G. Raisz and G.A. Rodan (Eds.), Principles of Bone Biology, Chapter 50, Interleukin-6-type cytokines and their receptors, Academic Press, San Diego, pp. 701-713.
80. Parfitt, A.M., Travers, R., Rauch, F. and Glorieux, F.H. (2000) Bone 27, 487-494.
81. Manolagas, S.C. (2000) Endocr.Rev. 21, 115-137.
82. Parfitt, A.M. (1992) Triangle 31, 99-110.
83. Robey, P.G. and Bianco, P. (1999) in: C.Rosen, J.Glowacki and J.P.Bilezikian (Eds.), The Aging Skeleton, Chapter 13, Cellular Mechanisms of Age-Related Bone Loss, Academic Press, San Diego, CA, pp. 145-157.
84. Dempster, D.W. (1989) J. Bone Miner. Res. 4, 137-141.
85. Kember, N.F. (1978) Cell Tissue Kinet. 11, 477-485.
86. Kember, N.F. (1971) Clin.Orthop. 76, 213-230.
87. Baron, J., Klein, K.O., Colli, M.J., Yanovski, J.A., Novosad, J.A., Bacher, J.D. and Cutler, G.B. (1994) Endocrinology 135, 1367-1371.
88. Parfitt, A.M. (2002) Bone (in Press).
89. Smith, E.P., Boyd, J., Frank, G.R., Takahashi, H., Cohen, R.M., Specker, B., Williams, T.C.,

Lubahn, D.B. and Korach, K.S. (1994) N.Engl.J.Med. 331, 1056-1061.

90. Morishima, A., Grumbach, M.M., Simpson, E.R., Fisher, C. and Qin, K. (1995) J. Clin. Endocrinol. Metab. 80, 3689-3698.

91. Tajima, Y., Yokose, S., Kawasaki, M. and Takuma, T. (1998) Histochem. J. 30, 467-472.

92. Kousteni, S., Bellido, T., Plotkin, L.I., O'Brien, C.A., Bodenner, D.L., Han, L., Han, K., DiGregorio, GB., Katzenellenbogen, J.A., Katzenellenbogen, B.S., Roberson, P.K., Weinstein, R.S., Jilka, R.L. and Manolagas, S.C. (2001) Cell 104, 719-730.

93. Jilka, R.L., Hangoc, G., Girasole, G., Passeri, G., Williams, D.C., Abrams, J.S., Boyce, B., Broxmeyer, H., and Manolagas, S.C. (1992) Science 257, 88-91.

94. Bellido, T., Jilka, R.L., Boyce, B.F., Girasole, G., Broxmeyer, H., Dalrymple, S.A., Murray, R. and Manolagas, S.C. (1995) J. Clin. Invest. 95, 2886-2895.

95. Jilka, R.L., Takahashi, K., Munshi, M., Williams, D.C., Roberson, P.K. and Manolagas, S.C. (1998) J. Clin. Invest. 101, 1942-1950.

96. Pacifici, R. (1998) Endocrinology 139, 2659-2661.

97. Mbalaviele, G., Jaiswal, N., Meng, A., Cheng, L.Z., Van den Bos, C. and Thiede, M. (1999) Endocrinology 140, 3736-3743.

98. Taylor, L.M., Turksen, K., Aubin, J.E. and Heersche, J.N.M. (1993) Endocrinology 133, 2292-2300.

99. Weinstein, R.S., Nicholas, R.W. and Manolagas, S.C. (2000) J. Clin. Endocrinol. Metab 85, 2907-2912.

100. Weinstein, R.S., Jilka, R.L., Parfitt, A.M. and Manolagas, S.C. (1998) J. Clin. Invest. 102, 274-282.

101. Cheng, S.L., Zhang, S.F., Mohan, S., Lecanda, F., Fausto, A., Hunt, A.H., Canalis, E. and Avioli, L.V. (1998) J. Cell Biochem. 71, 449-458.

102. Tobler, A., Meier, R., Seitz, M., Dewald, B., Baggiolini, M. and Fey, M.F. (1992) Blood 79, 45-51.

103. Quarto, R., Thomas, D. and Liang, C.T. (1995) Calcif. Tissue Int. 56, 123-129.

104. Bergman, R.J., Gazit, D., Kahn, A.J., Gruber, H., Mcdougall, S. and Hahn, T.J. (1996) J. Bone Miner. Res. 11, 568-577.

105. Kahn, A., Gibbons, R., Perkins, S. and Gazit, D. (1995) Clin.Orthop. 313, 69-75.

106. Mueller, S.M., Mizuno, S. and Glowacki, J. (1998) Bone 23, S536. (Abstract)

107. Nishida, S., Endo, N., Yamagiwa, H., Tanizawa, T. and Takahashi, H.E. (1999) J Bone Miner.Metab 17, 171-177.

108. D'Ippolito, G., Schiller, P.C., Ricordi, C., Roos, B.A. and Howard, G.A. (1999) J. Bone Miner. Res. 14, 1115-1122.

109. Oreffo, R.O., Bord, S. and Triffitt, J.T. (1998) Clin.Sci. (Colch.) 94, 549-555.

110. Moore, S.G. and Dawson, K.L. (1990) Radiology 175, 219-223.

111. Gimble, J.M., Robinson, C.E., Wu, X. and Kelley, K.A. (1996) Bone 19, 421-428.

112. Bianco, P. and Robey, P.G. (1999) J. Bone Miner. Res. 14, 336-341.

113. Spiteller, G. (1998) Chem. Phys. Lipids 95, 105-162.

114. Bolton-Smith, C., Woodward, M. and Tavendale, R. (1997) Eur. J. Clin. Nutr. 51, 619-624.

115. Jira, W., Spiteller, G. and Schramm, A. (1996) Chem.Phys.Lipids. 84, 165-173.

116. Nourooz-Zadeh, J. and Pereira, P. (1999) Ophthalmic Res. 31, 273-279.

117. Lecka-Czernik, B., Grant, D.F., Manolagas, S.C. and Jilka, R.L. (2000) J. Bone Miner. Res. 15, S372. (Abstract)

118. Mosca, J.D., Hendricks, J.K., Buyaner, D., Davis-Sproul, J., Chuang, L.C., Majumdar, M.K., Chopra, R., Barry, F., Murphy, M., Thiede, M.A., Junker, U., Rigg, R.J., Forestell, S.P., Bohnlein, E., Storb, R. and Sandmaier, B.M. (2000) Clin. Orthop. S71-S90

119. Liechty, K.W., MacKenzie, T.C., Shaaban, A.F., Radu, A., Moseley, A.M., Deans, R., Marshak, D.R.

and Flake, A.W. (2000) Nat. Med. 6, 1282-1286.

120. Anklesaria, P., FitzGerald, T.J., Kase, K., Ohara, A. and Greenberger, J.S. (1989) Blood 74, 1144-1151.

121. Anklesaria, P., Kase, K., Glowacki, J., Holland, C.A., Sakakeeny, M.A., Wright, J.A., FitzGerald, T.J., Lee, C.Y. and Greenberger, J.S. (1987) Proc. Natl. Acad. Sci. USA 84, 7681-7685.

122. Nilsson, S.K., Dooner, M.S., Weier, H.U., Frenkel, B., Lian, J.B., Stein, G.S. and Quesenberry, P.J. (1999) J. Exp. Med. 189, 729-734.

123. Hou, Z., Nguyen, Q., Frenkel, B., Nilsson, S.K., Milne, M., Van Wijnen, A.J., Stein, J.L., Quesenberry, P., Lian, J.B. and Stein, G.S. (1999) Proc. Natl. Acad. Sci. USA 96, 7294-7299.

124. Pereira, R.F., O'Hara, M.D., Laptev, A.V., Halford, K.W., Pollard, M.D., Class, R., Simon, D., Livezey, K. and Prockop, D.J. (1998) Proc. Natl. Acad. Sci. USA 95, 1142-1147.

125. Horwitz, E.M., Prockop, D.J., Gordon, P.L., Koo, W.W., Fitzpatrick, L.A., Neel, M.D., McCarville, M.E., Orchard, P.J., Pyeritz, R.E. and Brenner, M.K. (2001) Blood 97, 1227-1231.

126. Horwitz, E.M., Prockop, D.J., Fitzpatrick, L.A., Koo, W.W., Gordon, P.L., Neel, M., Sussman, M., Orchard, P., Marx, J.C., Pyeritz, R.E. and Brenner, M.K. (1999) Nat. Med. 5, 309-313.

127. Suva, L.J., Seedor, J.G., Endo, N., Quartuccio, H.A., Thompson, D.D., Bab, I. and Rodan, G.A. (1993) J. Bone Miner. Res. 8, 379-388.

CONTRIBUTOR ADDRESSES

Stem Cells: A Cellular Fountain of Youth
Edited by Mark P. Mattson and Gary Van Zant

KENNETH R. BOHELER, Ph.D., National Institute on Aging,
Laboratory of Cardiovascular Sciences, 5600 Nathan Shock Drive,
Baltimore, MD, 21224, USA
Phone: 410-558-8095, Fax: 410-558-8150
e-mail: bohelerk@grc.nia.nih.gov

JINGLI CAI, National Institute on Aging,
Laboratory of Neurosciences, 5600 Nathan Shock Drive,
Baltimore, MD, 21224, USA
Phone: 410-558-8609, Fax: 410-558-8465
e-mail: caiji@grc.nia.nih.gov

KAREN CHANDROSS, Ph.D., Aventis Pharmaceuticals, CNS Division,
MS K-203A, Route 202-206, Bridgewater, NJ 08807
Phone: 908-231-3991, Fax: 908-231-2413
e-mail: karen.chandross@aventis.com

AIWU CHENG, National Institute on Aging,
Laboratory of Neurosciences, 5600 Nathan Shock Drive,
Baltimore, MD, 21224, USA
Phone: 410-558-7152, Fax: 410-558-8465
e-mail: chengai@grc.nia.nih.gov

HARTMUT GEIGER, Departments of Internal Medicine and Physiology,
University of Kentucky Medical Center, 800 Rose St., Lexington,
Kentucky 40536-0093, USA

AMIELA GLOBERSON, Ph.D., Department of Immunology,
Weizmann Institute of Science, 304 Wolfson Building, 76100 Rehovot ISRAEL
Phone: 972-8-934-3437, Fax: 972-8-934-4141
e-mail: amiela.gloerson@weizmann.ac.il

NORMAN HAUGHEY, National Institute on Aging,
Laboratory of Neurosciences, 5600 Nathan Shock Drive,
Baltimore, MD, 21224, USA
Phone: 410-558-8694, Fax: 410-558-8465
e-mail: haugheyno@grc.nia.nih.gov

ROBERT L. JILKA, Ph.D., University of Arkansas Medical School Hospital,
Division of Endocrinology and Metabolism, Slot 587, 4301 W Markham St,
Slot 587, Little Rock, AR 72205
Phone: 501 686-7896, Fax: 501 686-8954
e-mail: rljilka@life.uams.edu

ROBERT L. JILKA, Ph.D., Division of Endocrinology and Metabolism,
Slot 587, University of Arkansas for Medical Sciences, 4301 W. Markham,
Little Rock, AR 72205
Phone: 501-686-7896, Fax: 501-686-8954
e-mail: rljilka@life.uams.edu

ERIN L. MANNING, Departments of Internal Medicine and Physiology,
University of Kentucky Medical Center, 800 Rose St., Lexington,
Kentucky 40536-0093, USA

EVA MEZEY, National Institutes of Health, National Institute of Neurological
Disorders and Stroke, BNP, Bethesda, MD 20892

MARK P. MATTSON, Laboratory of Neurosciences,
National Institute on Aging, 5600 Nathan Shock Drive, Baltimore, MD 21224
Phone: 410 558-8462, Fax: 410 558-8465
e-mail: mattsonm@grc.nia.nih.gov

MICHAEL PARFITT, Division of Endocrinology and Metabolism,
Center for Osteoporosis and Metabolic Bone Diseases,
Central Arkansas Veterans Healthcare System,
University of Arkansas for Medical Sciences, Little Rock, AR 72205

MAHENDRA RAO, Laboratory of Neurosciences,
National Institute on Aging Gerontology Research Center,
5600 Nathan Shock Drive, Baltimore, MD 21224

MAHENDRA RAO, Ph.D., NIA/GRC, Laboratory of Neurosciences,
5600 Nathan Shock Dr., Baltimore, MD 21224
Phone: 410-558-8204, Fax: 410-558-8465
e-mail: raoma@grc.nia.nih.gov

DR. MICHAEL A. RUDNICKI, Ottawa Hospital Research Institute,
Program in Molecular Genetics, 501 Smyth Road, Ottawa, Ontario, Canada, K1H 8L6
Phone: 613-739-6740, Fax: 613-739-8803
e-mail: mrudnicki@ottawahospital.on.ca

PATRICK SEALE, Department of Biology, McMaster University, 1280 Main Street
West, Hamilton, Ontario, Canada L8S 4K1
Ottawa Hospital Research Institute, Program in Molecular Genetics, 501 Smyth Road,
Ottawa, Ontario, Canada K1H 8L6

THOMAS E SMITHGALL, Ph.D.,
Department of Molecular Genetics & Biochemistry,
University of Pittsburgh School of Medicine,
E1240 Biomedical Science Tower, Pittsburgh, PA 15261
Phone: 412 648-9495, Fax: 412-624-1401
E-Mail: tsmithga@pitt.edu

GARY VAN ZANT, Departments of Internal Medicine and Physiology,
University of Kentucky Medical Center, 800 S. Rose St., Lexington, KY 40536-0093
Phone 859-323-5719, Fax 859- 257-7715
e-mail: gvzant1@pop.uky.edu

ANNA WOBUS, The Institute of Plant Genetics and Crop Plant Research,
In Vitro Differentiation Group, The Institute of Plant Genetics,
Gerontology Research Center and Crop Plant Research, Corrensstr. 3 D-06466
Gatersleben, Germany

Advances in
Cell Aging and Gerontology
Series Editor: Mark P. Mattson
URL: http://www.elsevier.nl/locate/series/acag

Aims and Scope:
Advances in Cell Aging and Gerontology (ACAG) is dedicated to providing timely review articles on prominent and emerging research in the area of molecular, cellular and organismal aspects of aging and age-related disease. The average human life expectancy continues to increase and, accordingly, the impact of the dysfunction and diseases associated with aging are becoming a major problem in our society. The field of aging research is rapidly becoming the niche of thousands of laboratories worldwide that encompass expertise ranging from genetics and evolution to molecular and cellular biology, biochemistry and behavior. ACAG consists of edited volumes that each critically review a major subject area within the realms of fundamental mechanisms of the aging process and age-related diseases such as cancer, cardiovascular disease, diabetes and neurodegenerative disorders. Particular emphasis is placed upon: the identification of new genes linked to the aging process and specific age-related diseases; the elucidation of cellular signal transduction pathways that promote or retard cellular aging; understanding the impact of diet and behavior on aging at the molecular and cellular levels; and the application of basic research to the development of lifespan extension and disease prevention strategies. ACAG will provide a valuable resource for scientists at all levels from graduate students to senior scientists and physicians.

Books Published:
1. P.S. Timiras, E.E. Bittar, *Some Aspects of the Aging Process,* 1996, 1-55938-631-2
2. M.P. Mattson, J.W. Geddes, *The Aging Brain,* 1997, 0-7623-0265-8
3. M.P. Mattson, *Genetic Aberrancies and Neurodegenerative Disorders,* 1999, 0-7623-0405-7
4. B.A. Gilchrest, V.A. Bohr, *The Role of DNA Damage and Repair in Cell Aging,* 2001, 0-444-50494-X
5. M.P. Mattson, S. Estus, V. Rangnekar, *Programmed Cell Death, Volume I,* 2001, 0-444-50493-1
6. M.P. Mattson, S. Estus, V. Rangnekar, *Programmed Cell Death, Volume II,* 2001, 0-444-50730-2
7. M.P. Mattson, *Interorganellar Signaling in Age-Related Disease,* 2001, 0-444-50495-8
8. M.P. Mattson, *Telomerase, Aging and Disease,* 2001, 0-444-50690-X
9. M.P. Mattson, *Stem Cells: A Cellular Fountain of Youth,* 2002, 0-444-50731-0
10. M.P. Mattson, *Calcium Homeostasis and Signalling,* 2002, 0-444-51135-0

Printed and bound by CPI Group (UK) Ltd, Croydon, CR0 4YY

03/10/2024

01040419-0019